HOW SOILS WORK

A Study Into the God-Plane Mutualism of Soils and Crops

by

Paul W. Syltie, Ph.D.

xulon
PRESS

Xulon Press
11350 Random Hills Road
Suite 800
Fairfax, VA 22030
(703) 279-6511
XulonPress.com

To order additional copies, call 1-866-909-BOOK (2665).

Preface

Why attempt this "impossible" task of trying to convey a realistic understanding of how soils work? The answer is simple: because it is our responsibility as stewards of the earth to comprehend the basic principles of this most essential supporter of human and animal kind. Unless we grasp how soils function beneath our feet — unseen but vibrantly alive and active — we cannot properly attend to the soil's needs as the vital sustainer of civilization.

Humankind is served by the soil. We all desperately and utterly depend upon the soil for our existence. All of us require food to exist, and the vast majority of that food comes from plants that grow in the soil. The cotton, ramie, and linen for our clothes come from plants that grow in soils Wood comes from forests, and animal fibers come from the sheep, goats, and llamas that depend on grass and other plants that grow from the soil.

Besides, nearly every other phase of life is inextricably tied to the soil. I am speaking here of the top few inches of the earth's crust in which plants anchor their roots and carry out a myriad of functions to produce the food, feed, and fiber that sustain civilizations over all the earth. We may discard most phases of our economy — steel production, computers, rail and automobile travel, communications, and many other industrial functions — but as long as food and fiber crops can grow we can still survive. The components of corn, wheat, millet, barley, rice, alfalfa, soybeans, tomatoes, melons, potatoes, pasture grasses, and relatively few other crops provide the sustenance for every person on earth. Moreover, the raw materials that are extracted from these crops provide prime ingredients for other uses: corn for sugars and alcohol, soybeans for oil and protein concentrates, flax for paint oils, trees for lumber and paper, and the list goes on and on.

In order to correctly understand soils and the correct approach to work with them, one must go directly to the Creator of all the elements that comprise matter: all the minerals, microorganisms, plants, animals, birds, fish, and people that have been fabricated from these elements. Without the correct foundation on which to build there can be no correct wisdom to understand how soils work and how to properly utilize them. It is impossible to garner the correct knowledge — the truth on this or any issue — without first fearing God, for "The fear of the Lord is the beginning of knowledge, but fools despise wisdom and instruction."[1] Again, "The fear of the Lord is the beginning of wisdom ...."[2] Therefore, to discover the truth of the workings of the soils beneath our feet we must dedicate ourselves to achieving Elohim's wisdom, knowledge, and understanding of the matter. He was the Creator of the entire scheme of soils and plants, and of mankind who survives upon their prosperity. It is our job to understand the principles of the proper functioning of soils and crops through understanding His principles — which guide the entire created world — through scientific revelations and personal observations.

When mankind was created on the sixth day of creation, he was given a mandate from Elohim:

"And Elohim said, Let us make man in our image, after our likeness, and let them have dominion over the fish of the seas, and over the fowl of the air, and over the cattle, and over all the earth, and over every creeping thing that creeps upon the earth." [3]

Elohim then blessed the people He made and commanded them to be fruitful, multiply, and replenish and subdue the earth ... and to have *dominion* over all living things. This word *dominion* is very interesting, for it means literally to "tread down, i.e. subjugate" (Hebrew *radah*).[4] It is important to know what this concept means, not because it defines how the soil works — that issue was decided by God from the beginning — but reveals how mankind is to work with the soil ... how he is to manage it.

Looking a little further into Genesis, we see that Adam, even before Eve was created, was placed within a garden in the land of Eden.[5] This was the most perfect environment imaginable for a person to live. The word *Eden* means "paradise or pleasantness", a place of great enjoyment and fulfillment.[6] Into this garden paradise Yahweh Elohim placed Adam to *dress* and *keep* it. Here are two more very interesting words. To *dress* (Hebrew *abad*) means to "work, serve, or till", while *keep* (Hebrew *shamar*) means "to hedge about, or guard, protect, or attend to." [7] Both of these words imply *service* in the work of attending to the needs of the Garden of Eden, which shows that *dominion*, as just discussed, means not a harsh, pummeling, exploitative approach to the creation but an uplifting, building, and serving approach to the animals, fish, birds, creeping things, and plants in the garden. This service approach is summarized by the word "love" in the word of God, the designed essence of human existence on earth. It is the central core of Elohim's design for each person to live in harmony with his fellow man ... the entire meaning of the law and the prophets:

"Therefore all things whatsoever you would that men should do to you, do you even so to them, for this is [the meaning or intent of] the law and the prophets." [8]

This selfless approach towards life is the inherent character of the ten commandments and the statutes and judgements, which are summarized as follows:

"You shall love the Lord your God with all your heart, and with all your soul, and with all your mind. This is the first and great commandment, and the second is like it: You shall love your neighbor as yourself. On these two commandments hang all the law and the prophets." [9]

We as inhabitants of the earth are, on the other hand, in great conflict with nature. Rather than love and service being expressed amongst creatures around us — in both the plant and animal world — we see competition, strife, predation, and suffering. Elm trees are attacked by bark beetles which spread deadly Dutch elm disease; fungal blights of various races attack potatoes, bananas, and many other crops; army worms strip pastures and grain fields; grizzly bears prey on elk and deer, while wolves and coyotes hunt and consume rabbits and moose; lions attack the young and old of harmless antelope and wildebeests. The list is endless of fierce competitiveness in the natural world.

This competition and strife, disease and death is not "normal"... or should we say, the intended order of things as instigated by the Creator of all things. It is a consequence of mankind allowing an insidious, subtle spirit to come into play within the creation. We know this spirit as Satan or Lucifer.[10] He has tainted all of this world with an attitude of competition and self-seeking ... of the strong preying on the young, weak, or infirm.[8] This attitude of self-centeredness leads to the disease and death that is evident in the natural world all around us, for rather than support the weak and suffering the current tenet of nature is to destroy and consume them, remove them from the scene, and forget about compassion and kindness. This "law of the jungle" is well-stated in I Peter 5:8:

"Be sober, be vigilant, because your adversary the devil, as a roaring lion, walks about seeking whom he may devour."

There are, then, the two essential ways of living that are extant within all of nature around us:

The Way of Life	The Way of Death
Aggradation	Degradation
Well-being	Suffering
Abundant life	Disease and death
Rejoicing	Mourning
Peace	Disruption and war

The contrast between these two basic ways of existence are in evidence at every level of the universe: the plant and animal kingdoms, man and his culture, elements of the weather ... even to the outer reaches of the cosmos. It is this conflict that has frustrated man's attempts to achieve happiness and permanence since the Garden of Eden ... where for a short time truly abundant life, joy, and peace existed within the plant and animal kingdoms, and within the lives of the two people living in that garden.

It is essential to understand these two basic forces existing side-by-side in nature to understand how soils work. There exists an ideal system whereby no forces of evil impinge upon the soil and plants.[11] None of us have ever experienced such a situation, at least for very long. Eventually, evil creeps in and attempts to wreak its destruction, but fortunately powerful defense systems are built into all living things that aid them in resisting and overcoming these destructive forces. Human and animal bodies have highly effective immune systems, as do plants in their own way. Allies within the microbial, plant, or animal world come into play, attacking the attackers. We call them beneficial insects, friendly bacteria and fungi, or hydrogen peroxide in rainwater to weaken the attack force. Sometimes unseen forces do the defending, like the spiritually motivated flies, frogs, hail, boils, and death of the firstborn that overthrew the powerful Egyptian dynasty in Moses' time.[10]

To understand how soils work we must first of all see how water, air, minerals, organic matter, macro and microorganisms, and plant roots are designed to work. Then we must understand how those perfectly ordered systems are afflicted by the destroyers within an imperfect world. Remember that this world and its integrated, beautifully harmonious systems — both natural and man-made — prosper according to the degree that the designed functions can proceed unimpeded by destructive elements. Throughout this study we will see interwoven the two systems of life: give and get, service and self-centeredness. This is why the first lesson to be covered will deal with *mutualism*. This concept, well-known in biology, forms the heart of understanding how soils work. It is really the biologist's means of defining the systems based on love, and those based on greed.

There will be no attempt to extensively cover issues of soils that are addressed in texts such as *The Nature and Properties of Soils,* although frequent reference to facts in this soils text and in other soils and ecology texts will be made. One can consult many excellent outside sources for more detail if desired. The essentials for understanding various topics will be covered, and the reader is encouraged to dig more deeply into any particular topic of interest. May this guide on how soils work prove to be a most illuminating, eye-opening adventure for you as we work together to uncover the truth that lies beneath our feet!

Dedication

This book on soils and plants is dedicated to the Great Creator God who made all things, and by whose power all things exist. It has been the objective of this work to extend but a very humble effort to bring alive a small fraction of the wonderful, extensive truths concerning the soils and crops that feed us ... how His express character has been integrated into every facet of the living and nonliving things around us. They all form a part of the mutualistic whole based upon lovingkindness.

This is not to say there has not been considerable infection of the Creation with the Adversary's genetics and character ... but even those evils help us face the reality of how good and great, in contrast, are the things of the One who in the beginning made all things perfect and at peace. What a wonderful Creator we have to provide so well for us our physical necessities of life through the complex processes of the ecosphere that He awesomely fashioned for us, and to allow the truths of this Creation to come alive to those who will but seek to understand them.

Likewise, how wonderful a Creator we have who has set the choice between good and evil squarely before us in all that He has made, messages and answers that cannot be suppressed or hidden despite the ravages of humanistic education.

TABLE OF CONTENTS

List of Figures

Chapter 3

Chapter 4

List of Tables

We Must Look Ahead

Soil management based on conventional tillage, pesticides, herbicides, and heavy machinery has produced satisfactory — even abundant — yields for many years in Western agriculture. Yet, these practices are rooted in self-centered exploitation and are inherently self-destructive in the long term. They will not produce high yields forever, but will lead to a gradual reduction in the soil's ability to sustain high yields and high crop quality. Industrial agriculture may produce high yields, but yields of questionable quality that are totally dependent on the continued use of petrochemicals and pesticides ... which may be disrupted. Depletion of soil minerals and organic matter, and destruction of tilth and strong structure, are hard to avoid when exploitative industrial tenets are applied to biological systems in soils.

What works short-term may not work long-term if the practices used on our land are insidiously degenerative. Witness the devastation of soil depletion and erosion to the old Cotton Belt of the South, the ravages of Roman exploitation in the wheat belt of North Africa, and the denuding of Palestine's and Greece's fields and forests in the face of overcultivation and goat grazing.

We must determine what works long-term to build and maintain soil health, and also take into account effects of these practices on the social, spiritual, economic, and ecological issues of the present and future. What effect does having only 2% of the country's population producing food — and then under considerable financial duress — have on society as a whole? Can a depressed farm sector be expected to rebuild soil fertility, or exploit and diminish it? Will food quality be minimized when varieties and cultural practices are designed to squeeze out high yields that contain low nutrient levels? Can a farmer pay attention to the unique needs of an erosive hill or corner of a field when time, labor, and debt considerations force him to speed his way across his 3,000-acre spread?

Industrial economics dictate that the farmer maximize short-term gain utilizing economies of scale that encourage ever-larger acreages, gargantuan machinery, increasing debt, reduced labor, and expensive but bulk-producing crop varieties and chemicals. Today's game of soil management forces short-term views if the farmer buys into the industrial paradigms encouraged to replace the family farm.

We must look long-term at soil and farm issues. We must look to the source of our life —- the Creator of ourselves and the soil, and the systems He designed expressly to sustain mankind — and fear Him, not the banker. If our choices in soil management are not geared towards "Doing to others as we would have others do to us," how can the future hold any promise? Can we reap joy in daily living on the land, and increase the joy of our neighbor, if we forget to insure future prosperity through the activities we accomplish today?

HOW SOILS WORK

A Study Into the God-Plane
Mutualism of Soils and Crops

by

Paul W. Syltie, Ph.D.

Chapter I
Mutualism: the Key to Healthy Plants and Soils

This is really a story about not only how soils work, but how all things work ... at least work in the best possible way. There are absolutes that pervade all of nature and the entire universe. One way — the way of uprightness and lawful living — yields abundance and joy, while the other way — of selfishness and anarchy — produces eventual bankruptcy and despondency.

These two approaches are evident throughout the earth and the entire cosmos. We see the sun in its pure brilliance, radiating its full-spectrum energy throughout the solar system, nourishing and warming all life on earth ... making our existence possible. A trillion stars in the heavens likewise twinkle at us in comforting assurance that there is indeed a way of goodness and truth. Then there are tornadoes outbreaks, locust attacks, bee stings, and scratches from thorn bushes that remind us all things are not perfect. Perhaps the potato plants froze last night in a late spring cold snap, or a flood eroded the driveway, muddying the nearby creek. Things are not all good in this world ... but they could be if the Author of all good was allowing no evil to prosper.

However, He is not preventing that evil today.[1] While the Creator made all things good in their original orders[2], a vicious and powerful being has been very busy polluting, confusing, and disrupting the perfect order designed originally in Eden. This Lucifer, also called the devil, Satan, and Destroyer[3] is the present "god of this world" [4], the one who has perverted the perfection of the created world and the beautiful processes designed to produce peace and abundance for everyone on the earth. Lions, originally grass eaters, are now predacious flesh eaters.[5] Thorns and thistles, which cause pain to protect themselves, were not a part of the original Edenic creation.[6] Neither were ticks, mosquitoes, and horse flies which thrive on mammalian blood. Truly, the beautiful, painless, disease-free order of the original creation that was "very good" [7] is no longer with us. It is tainted with evil.

Mutualism — the Key to How Soils Work

A key to understanding how the soil functions is to understand the concept of *symbiosis*. Symbiosis is defined as "two kinds of organisms living together to the advantage of at least one organism".[8] The organisms in the relationship are called *symbionts*.

Symbiosis comes from the Greek *symbioun*, meaning "to live together".[9] Not always is this association beneficial to both, as shown in the following types of symbiosis.[10]

1. Mutualism. Both organisms benefit from the association. This can be a plant with a plant, a plant with an animal, or an animal with an animal. Examples:

Lichen. Two plants live together. A fungus layers itself against a rock or wood and extracts nutrients, supplying some of its nutrients to a layer of algae atop it, which in turn provides photosynthate for itself and the fungus. The algae also gains protection and a greater water supply from the fungus. Together, they can live in a hostile environment where neither could live alone.

Rhizobium bacteria. These bacteria form nodules on legume roots and are thus protected while fixing atmospheric nitrogen. The plant provides photosynthate for the bacteria and gains nitrogen compounds essential for growth. See Figure 1-1 for a photograph of Rhizobium bacteria.[11]

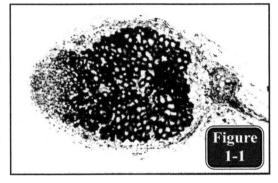

Figure 1-1

Bees pollinating flowers. Some flowers require a bee or other insect to wade through its stamens (pollen-bearing organs) and deposit pollen on the pistil. In the process, the bee gets nectar and pollen, and the flower becomes fertilized for seed production.

Water buffalo and tick bird. Tick birds eat ticks from the water buffalo, getting food while the animal gets rid of ticks (Figure 1-2).[12] The birds also warn of danger by uttering shrill cries and jumping up and down.

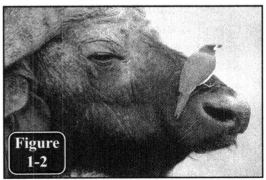

Figure 1-2

Facultative mutualism: Either organism can live without the other, such as the rhinoceros and tick bird.

Obligatory mutualism: Neither organism can live without the other. For example, termites require certain protozoa to live in their intestines.

These protozoa possess enzymes that digest wood cellulose, without which the termites would die. Antibiotics that kill the protozoa also kill the termites due to starvation. The protozoa cannot survive outside the termite's intestines.

2. Commensalism. One symbiont benefits from the association and the other is neither benefited nor harmed. Examples:

Coral reef anemone fish and sea anemones. The coral reef anemone fish nestles among the tentacles of the giant sea anemone for protection from predator fish, which avoid the poisonous stings of this sea creature. The fish benefit but the anemone gets no benefit in return (Figure 1-3).[13]

Figure 1-3

Green hydra and algae. Small one-celled algae live within the gastro-dermal cells of the hydra, but do not benefit or harm the hydra. It can live just as well without them.

People and mouth bacteria. It is thought that bacteria living within the mouth are opportunistic and cause neither help nor harm to humans.

3. Parasitism. One symbiont benefits while the other is harmed. Examples:

Tapeworms and vertebrate animals. The tapeworm extracts food from the animal and excretes toxic waste products, harming the animal.

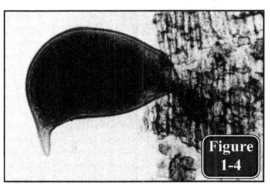

Figure 1-4

Leeches and animals or fish. These sucking creatures attach themselves to animals or fish and extract blood, harming or even killing the host.

Sucking nematodes and plants. Several species of nematodes that possess a stylet (spear-like mouth for sucking) puncture cells, inject enzymes, and extract nutrients, usually killing the cell and harming the host. Among the more common types are *Meloidogyne* (root-knot), *Heterodera* (cyst), Reniform (citrus), and *Pratylenchus* (lesion) species. See Figure 1-4 for a typical root-feeding nematode.[14]

Mutualism and God's Design

It is immediately apparent that *mutualism* forms the heart of how soils are designed to work, for both participants in the association are benefitted. This is the law of love in operation, not only in soils but throughout the universe. The opposite of mutualism is parasitism, the type of symbiosis where one creature benefits while the other suffers. Thus, we can decipher which organisms around us are untainted descendants of God's original loving creation that is defined as *good*, and which organisms have been altered by Satan to be self-serving, predacious, and hurtful.

Mutualistic organisms ⟶ Beneficial to living systems ⟶ Loving, selfless, cooperative, constructive, serving of others, lawful

Parasitic organisms ⟶ Detrimental to living systems ⟶ Conflicting, predacious, self-serving, violent, destructive, lawless

---

We must add to the mutualistic organisms mentioned above those organisms that attack parasites and defend the mutualistic organisms. These may be commensal in character on most occasions, or even mutualistic, but will rise to the occasion if the host is attacked ... an external immune system, as it were. An example would be predacious nematodes that roam the soil environment seeking other nematodes to eat (Figure 1-5).[15] They possess a stylet, but usually attack only those nematodes that are relatively abundant, such as during an outbreak of *Meloidogyne* or other pathogenic nematodes. Tiny soil mites will

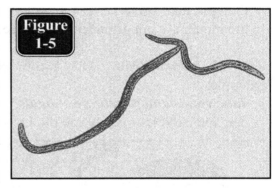

Figure 1-5

eat nematodes that attack plant roots, gobbling up dozens or hundreds each day. They are not symbiotic per se, but help defend roots like white blood cells eat bacteria and foreign material in the blood and other tissues of the human body. This army of defenders in the root zone may be said to be an "extended immune system" that reaches out beyond the boundaries of the root cells, like a country's army strives to repel foreign troops trying to invade and conquer (Figure 1-6).[16]

Mutualism indeed embodies the essence of the Creator's design. A host of scriptures support this view, a few of which are written below.

> "Bring you all the tithes into the storehouse, that there may be meat in My house, and prove Me now herewith, says the Lord of hosts, if I will not open you the windows of heaven and pour you out a blessing, that there shall not be room enough to receive it."[17]

> "And let us not be weary in well doing, for in due season we shall reap if we faint not. As we have opportunity, let us do good unto all men, especially unto them who are of

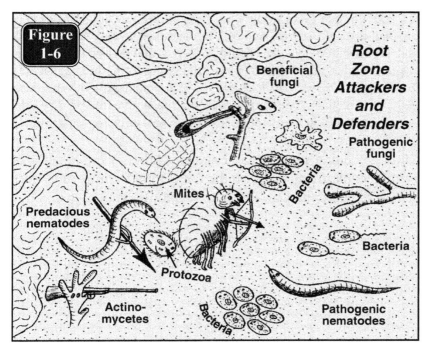

Figure 1-6

Root Zone Attackers and Defenders

Beneficial fungi

Pathogenic fungi

Bacteria

Mites

Predacious nematodes

Bacteria

Protozoa

Actinomycetes

Bacteria

Pathogenic nematodes

the household of faith."[18]

"He that has pity upon the poor lends unto the Lord, and that which he has given will he pay him again."[19]

"Cast your bread upon the waters, for you shall find it after many days."[20]

"But this I say, He which sows sparingly shall reap also sparingly, and he which sows bountifully shall reap also bountifully .... And God is able to make all grace abound toward you, that you, always having all sufficiency in all things, may abound to every good work ...."[21]

"And whosoever shall give to drink unto one of these little ones a cup of cold water only in the name of a disciple, truly I say unto you, he shall in no wise lose his reward."[22]

"... but you shall open your hand wide unto him [a poor countryman], and shall surely lend him sufficient for his needs, in that which he wants."[23]

"... the righteous gives and spares not."[24]

"Ask and it shall be given to you, seek and you shall find, knock and it shall be opened unto you; for everyone who asks receives, and he that seeks finds, and to him that knocks it shall be opened. Or what man is there of you whom, if his son asks for bread, will give him a stone? Or if he asks for a fish, will give him a serpent? If you then, being evil, know how to give good gifts to your children, how much more shall your Father which is in heaven give good things to them that ask Him? Therefore all things whatsoever you would that men should do to you, do you even so to them, for this is the law and the prophets."[25]

"Love your enemies, do good to them that hate you, bless them that curse you, and pray for them that despitefully use you. And unto him that smites you on the cheek, offer also the other, and him that takes away your cloak forbid not to take your coat also. Give to every man that asks of you, and of him that takes away your goods ask them not again. And as you would that men should do to you, do you also to them likewise .... But love your enemies, and do good, and lend, hoping for nothing again, and your reward shall be great .... Give and it shall be given unto you, good measure, pressed down, and shaken together, and running over, shall men give into your bosom. For with the same measure that you mete withal it shall be measured to you again."[26]

The law of love, of doing good to God and neighbor, is the essence of mutualism. This law and its ramifications form the foundation for how soils work ... or should be say, how they are intended by their Creator to work.

Chapter II
The Structure and Function of the Plant-Soil System

A Unity Among All Systems and Cycles

It is impossible to separate the structure (makeup) of a living organism from its function (how it works) ... for life implies activity, motion, and process. Likewise, it is impossible to separate the plant from the soil, since each depends on the other. That fact will become more apparent as this discussion continues.

One must have at least a vestigial understanding of the structure and functions of plants and their interrelationships with the environment to understand root function within the soil.

Figure 2-1[1] illustrates the cycling of water, air, and minerals. Cycling by definition implies *service*, service of one step giving to the next[2].

The Hydrologic (Water) Cycle. Atmospheric water vapor condenses to produce rain, which nourishes the soil, granting water essential for life processes of soil organisms and associated plants. Water is taken up by plant roots to perform its many functions within the roots and tops. These functions include ...

- Universal solvent for biochemical reactions
- Transporter of nutrients (up through the xylem) and photosynthate (down through the phloem)
- Coolant during evaporation from leaves (latent heat of vaporization)

Water evaporates from leaf and soil surfaces, as well as lakes, rivers, and oceans, and is moved by air masses over land where precipitation again nourishes the soil and root systems.

The Gas Cycle. Air is comprised of about 80% nitrogen (N_2), 29% oxygen (O_2), 0.03% carbon dioxide (CO_2), and small amounts of water vapor (H_2O), rare gases, and other compounds such as oxides of nitrogen, sulfur, and carbon. Oxygen exchanges with leaves during photosynthesis, and CO_2 is taken up and incorporated into carbohydrates and other plant compounds during photosynthesis. Some of these compounds are moved into the root zone for building more roots, or exuded into the soil alongside the roots to feed bacteria and fungi.

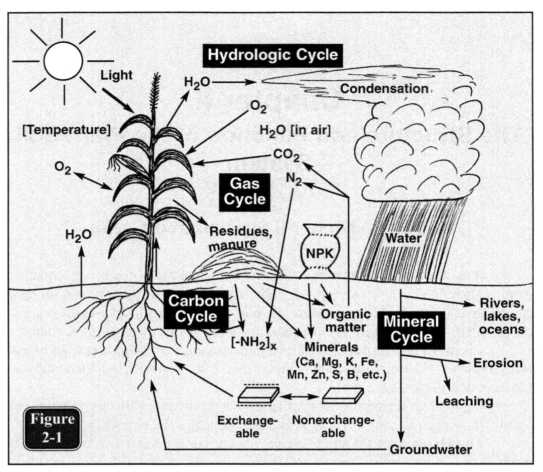

Figure 2-1

Oxygen, CO_2, and N_2 exchange from the air into soil pores to provide roots with adequate O_2 for biochemical reactions and thus growth, leading to root extension. Nitrogen in the soil is fixed by certain nitrogen-fixing organisms, either photosynthetic free-living autotrophs (as cyanobacteria and *Azospirillum*), free-living heterotrophs (as *Azotobacter*), or symbiotic types (as Rhizobium in legume nodules). The dinitrogen is reduced in oxidation state and incorporated into organic compounds (-NH_2). Some nitrogen is fixed by cyanobacteria on leaf surfaces, or even within the leaf stomata, under certain conditions. On plant death, nitrogen is broken down from tissue residues and returns to organic matter for another generation of plants.

The Carbon Cycle. This cycle is involved with the gas cycle, since CO_2, the oxidized form of carbon, is a gas. Carbon dioxide in the air is taken in by leaves and fixed as carbohydrates and other storage or structural compounds during photosynthesis and also moved into the root zone. With root and top death, carbon compounds are broken down to CO_2 once again or reworked into the highly complex lignaceous compounds of organic matter. These organic compounds, some easily broken and some resistant to breakdown for many years, supply N, S, and various minerals and complex organic stimulants such as humic and fulvic acids to stimulate plant growth. Besides, the organic matter has a very high nutrient-holding capacity (called *cation exchange capacity*), which holds K, Ca, Mg, and other positively charged ions until replaced by other cations in mass action reactions or by root exchange. Carbon dioxide is released to the air from organic matter breakdown over time and is used by leaves for reincorporation into living tissues.

      The Mineral Cycle. Each element of the mineral kingdom has its own unique cycle, but there are common characteristics to them all. The soil releases its stored elements — from exchangeable or non-exchangeable sites on clay or organic matter — to root hairs or to microorganisms (such as mycorrhizae) that extract the nutrients and move them to the roots. They are then taken up into the stems and leaves and utilized for various metabolic processes. In particular, the micronutrients Zn, Cu, Fe, and Mn act in part as enzyme cofactors (to make enzymes work), while Ca is a major component of cell walls and many metabolic pathways. Sulfur is an essential component of some amino acids, which are protein building blocks, and of vitamins, also essential to activate many enzymes. Boron is involved in sugar manufacture and transport, while phosphorus and its compounds are essential for energy capture and storage. Magnesium comprises the core of chlorophyll, the light energy trapping compound. These and other elements and their functions will be discussed later in more depth.

      Once the jobs of these elements are complete and the plant tissues die, microbes decompose the tissues and return the minerals to the soil for recycling to succeeding plant generations. Some of the elements may be transported out of the system by crop removal, leaching, or erosion, but they are replenished by continual weathering of soil minerals, and fertilizer and organic matter additions to the soil.

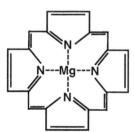

Chlorophyll nucleus [3]

      Note that the sun and its energy frequencies sent to earth are not cyclical. Sunlight is the one component of the system that is given freely and without any return ... signifying the totally selfless nature of the Creator who set the entire plan into motion. Without this input, the earth would quickly become a lifeless orb, bereft of any activity ... a frozen hulk with no function but to wander pointlessly in space. We owe our physical existence to that one basic reaction of sunlight capture by chlorophyll, and that energy's entrapment into energy-rich compounds which in turn power chemical reactions which produce carbohydrates, proteins, nucleic acids, lipids, vitamins, regulators, and a myriad of other compounds that comprise physical life. While this is not a biochemistry course, the essential steps of photosynthesis are shown in Figure 2-2. Each

11

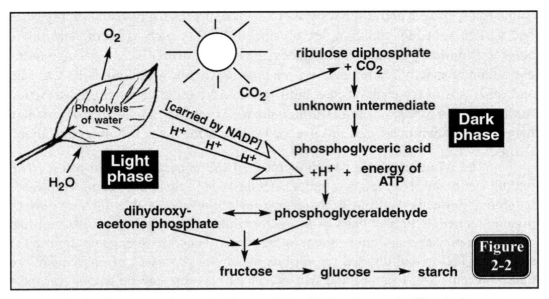

Figure 2-2

step of the cycle serves the next phase of the cycle, the total effort being to serve and provide for the growing plant so it may complete its germination, vegetative growth, and finally its reproductive growth.

The Plant and Its Growth

Plants are marvelous organisms that are designed to serve mankind, giving him food, fiber, and feed. It is necessary to cover at least a few vestigial basic concepts of plants and their operation so we may understand soils ant their functions later on. All of this discussion is being done at the risk of oversimplifying a profoundly complex organization within plants.

Tissues and Cells Common to All Plants

All plants have certain things in common: roots, stems, and leaves. All of these plant parts are comprised of cells, microscopic living structural and functional units which make up all living things, from a single-celled bacterium to a mammoth blue whale or sequoia tree. The roots, stems, and leaves in turn are composed of different cellular types for specialized functions such as structural support, energy storage, transport, protection, and reproduction. All cells and tissues are derived from meristems, the actively dividing portions of roots and stems, and may be divided into simple or complex tissues.

Simple Tissue (Figure 2-3)[5,6]

Parenchyma — unspecialized tissue that makes up a large portion of fruits, roots, and tubers and is involved with storage, photosynthesis, assimilation, and secretion.

Collenchyma — elongated cells that occur in strands or cylinders and are primarily for support in early growth.

Sclerenchyma — thick-walled cells having small cavities, in various configurations: single or massive. They are heavily lignified and are nonliving when mature. They may be long and fibrous and can interlace and form sheets. Clusters of irregularly shaped cells (scleroids) form hard shells, like those of pecans or fruit pits.

Complex Tissue — *combinations of simple and specialized cells and tissues*

The two major types of complex tissues in plants are xylem and phloem (Figure 2-4)[7].

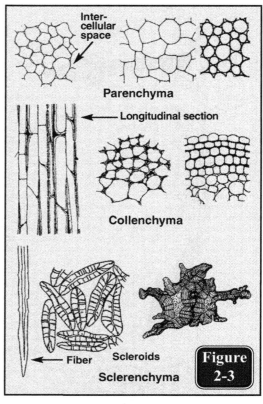

Figure 2-3

Photosynthesis and Energy Capture

Photosynthesis, which has already been briefly described, involves trapping of sunlight energy into the bonds of carbon compounds. This energy economy is briefly illustrated in Figure 2-5[8,9]. Note how sunlight energy is trapped through biosynthetic processes (involving chlorophyll and photosynthesis), utilizing water, nitrogen, and minerals from the root vascular stream and CO_2 from the air to yield the carbohydrates, proteins, lipids, and other compounds that are essential for plant structural organs and energy reserves. Respiration utilizes the energy stores for osmotic, electrical, mechanical, and chemical work of the cells, while various compounds are transferred by the phloem to other plant locations.

An actual picture of the leaf structures in which the biosynthetic processes of Figure 2-5 occur is shown in Figure 2-6[10]. The reactions occur primarily in the parenchyma cells. Some plants are "C-4" plants, containing enzymatic pathways in the leaf mesophyll cells that allow for more efficient carbon fixation under some conditions. Corn, sorghum, and sugar cane are such plants. Other plants have just a "C-3" pathway, as shown in Figure 2-7[11]. This pathway operates in the bundle sheath cells of the plant.

Once photosynthesis has occurred and energy has been stored in carbon compounds, this energy is utilized for life processes. While the biochemistry is

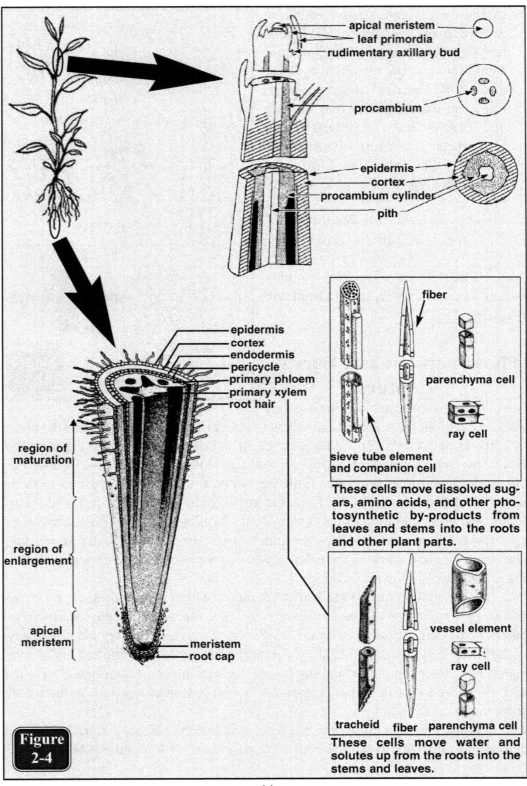

apical meristem
leaf primordia
rudimentary axillary bud

procambium

epidermis
cortex
procambium cylinder
pith

epidermis
cortex
endodermis
pericycle
primary phloem
primary xylem
root hair

region of maturation

region of enlargement

apical meristem

meristem
root cap

fiber

parenchyma cell

ray cell

sieve tube element and companion cell

These cells move dissolved sugars, amino acids, and other photosynthetic by-products from leaves and stems into the roots and other plant parts.

vessel element

ray cell

tracheid   fiber   parenchyma cell

These cells move water and solutes up from the roots into the stems and leaves.

Figure 2-4

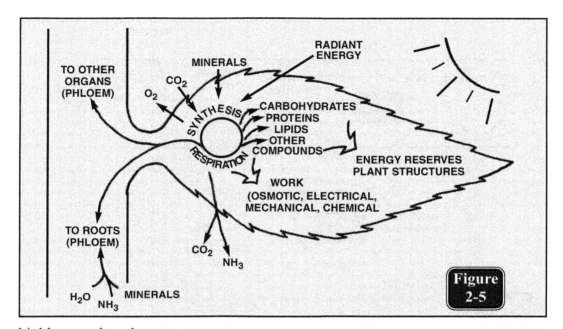

Figure 2-5

highly complex, the summary of the major reactions shown in Figure 2-8 is an overview of the essentials of the chief respiratory pathways of plants.[33] The essential portions of the entire integrated, mutualistic systems involves glycolysis, the Krebs cycle, and attendant electron transport. Each step along the intricate pathways serves a supply of substrate and energy for the next step,

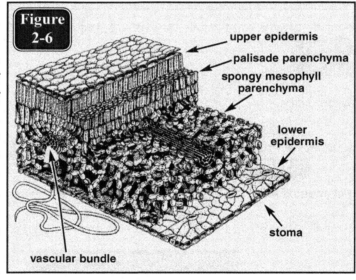

Figure 2-6

with enzymes expediting the handoff, as it were[12].

Throughout all of the cyclical pathways, the selfless gift of light from the sun is captured by chlorophyll and shunted into various compounds. The movement of this energy is on a service basis: each compound depends on its survival by the gift of energy passed on from its precursor, and in turn the next compound of the chain owes its existence to the one before it. Ultimately, the many cyclical processes sequestered within the cells of every tissue of the plant are powered by mutualistic efforts that gain their initial power thrust from the sun.

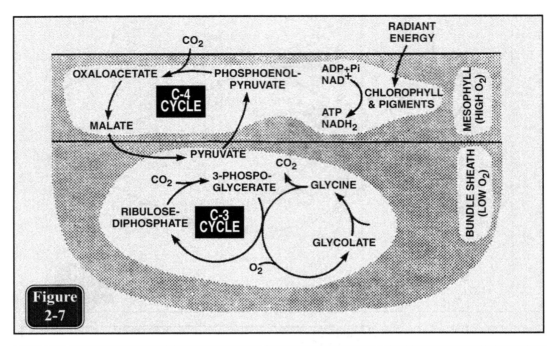

Figure 2-7

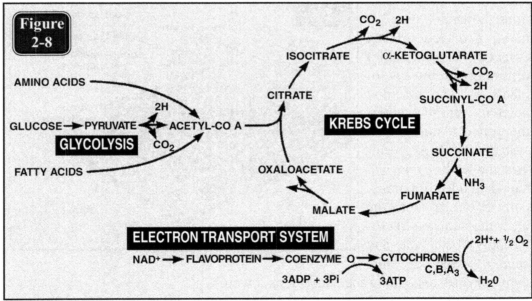

Figure 2-8

The Soil

The solid *terra firma* beneath our feet holds many wonderful secrets that are only now beginning to be unveiled. Studying the soil, as close as it is to us, has been likened to exploring the mysterious ocean depths ... so elusive have its secrets been.

In order to understand how the soil works, we will first view the essential *composition* of the soil, and then view its *functions*. All soils differ from one location to another across the planet, based upon subtle or major differences in climate, topography, parent material, vegetation, and the time these forces have exerted their effects. These are the so-called "factors of soil formation" that C.E. Kellogg and Hans Jenny outlined decades ago[13].

Although many men have tried to classify soils, these efforts have been only partially successful because no two soils are alike. Classification of soils is like classification of people: some are white, some are black, some are red, and some are brown, with varying ranges of limb and trunk proportions, cranial capacity and dimensions, musculation, nose and lip width, and so forth. Soils vary in color (related to organic matter and mineral contents), depth of various horizons, mineralogy at different depths, structure and composition of horizons, microorganism content, clay types, and so forth. They are about as fascinating in their differences as are people.

Soil Minerals and Elements

The soil is made up of particles that vary in size from extremely small — less than microscopic size — to rock or gravel. These particles may be arbitrarily categorized as sand, silt, and clay based on their sizes, and then grouped in textural classes depending on the

Soil separate textures	Diameter range (mm)
Very coarse sand	2.0-1.0
Coarse sand	1.0-0.5
Medium sand	0.5-0.25
Fine sand	0.25-0.10
Very fine sand	0.10-0.05
Silt	0.05-0.002
Clay	< 0.002

Table 2-1

proportion of each particle size (see Table 2-1)[14]. This categorization gives rise to the "texture triangle" seen in so many soils texts (Figure 2-9)[15].

These various particles are made up of minerals that vary with the particle size. The sand fraction tends to be highest in silicon, while finer particles are higher in iron and aluminum oxides. This varia-

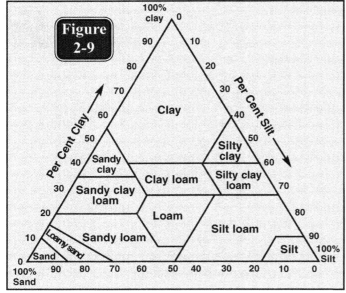

Figure 2-9

tion in size fraction composition is emphasized by the composition of a typical agricultural soil in Table 2-2[16].

Chemical Composition of a Montalto Silt Loam

Size of Particles	SiO_2	Fe_2O_3	Al_2O_3	TiO_2	CaO	MgO	
				%			
Sand	86.3	5.19	6.77	1.05	0.37	1.02	Table 2-2
Coarse Silt	81.3	3.11	7.21	1.05	0.41	0.82	
Fine Silt	64.0	9.42	12.00	1.05	0.32	2.22	
Coarse Clay	45.1	13.50	21.10	0.96	0.38	2.09	
Fine Clay	30.2	17.10	22.80	0.88	0.08	1.77	

As a whole, mineral soils are comprised of about 90% oxygen, silicon, and aluminum; organic soils contain mostly oxygen and carbon. Of these elements, only silicon plays a proven nutrient role in plants. Iron, the fourth most common element in the earth's crust, is used in very small amounts by plants, so in terms of chemical composition the essential nutrients that plants remove from the soil originate from a small percentage of the soil[17].

Of the elements in the soil, only a very small amount is taken up annually by plants. Thus, the soil can be viewed as a vast storage reservoir of nutrients, as revealed by Table 2-3[18].

Nutrient	Soil content	Annual Plant Uptake	Years of supply in a 4-inch soil layer	
	% by weight	lb/acre	years	
Calcium	1	45	260	
Potassium	1	27	430	Table 2-3
Nitrogen	0.1	27	50	
Phosphorus	0.08	6	150	
Magnesium	0.6	3.6	2,000	
Sulfur	0.05	1.8	320	
Iron	4.0	0.5	100,000	
Manganese	0.08	0.4	3,000	
Zinc	0.005	0.3	2,000	
Copper	0.002	0.1	1,000	
Chlorine	0.01	0.05	200	
Boron	0.001	0.03	400	
Molybdenum	0.0003	0.003	1,000	

The differences in chemical composition of the different particle sizes reflect the basic mineralogy of the sizes. Sand tends to be made up primarily of SiO_2, whereas the fine particles are comprised of clay minerals. In most soils, these minerals are fairly stable in the state they are found, having been derived from rocks that physical, chemical and microbial forces have degraded, or that water or wind have deposited some time in the past. The processes of soil formation are active on a continuing basis to gradually transform the mineralogy and particle sizes until they are relatively stable with the environment. However, the long-term effect of these forces of nature, especially microbial activity, is to degrade particles and release minerals so they can be leached or eroded, and thus lost to the system. On the other hand, plants and their roots tend to capture and cycle soil nutrients so they remain with the system. An excellent example of this recycling of nutrients within a very mineral-deficient soil system is a tropical rain-forest soil, where the majority of minerals are always in the trees and undergrowth, and the very weathered, leached, and depleted soils remain very low in nutrients. Leaves that fall are rapidly degraded, and the roots quickly absorb their nutrients.

The parent material greatly affects the mineral composition of the particle sizes, as one might expect. Thus, a soil developed on a high calcium limestone would logically contain more calcium in all of its separates than a highly leached soil developed on a coastal plain. Notice the values in Table 2-4[19].

Mineral content of soils from different parent materials

Particle size	Crystalline rock	Limestone	Coastal plain sand	Glacial till or loess*
Table 2-4				
		% phosphorus		
Sand	0.03	0.12	0.03	0.07
Silt	0.10	0.10	0.10	0.10
Clay	0.30	0.16	0.34	0.38
		% potassium		
Sand	1.33	1.21	0.31	1.43
Silt	2.00	1.52	1.10	2.00
Clay	2.37	2.17	1.34	2.55
		% calcium		
Sand	0.36	8.75	0.05	0.91
Silt	0.59	7.83	0.14	0.93
Clay	0.67	7.08	0.39	1.92

*Till means assorted materials deposited by glaciers. Loess means windblown silt and clay deposited on a soil surface.

Contained within the soil particles is a wide variety of elements. In some cases nearly every one of the elements on the periodic table can be found, if only in minute traces. One authoritative source gives the following documented contents of soil elements in surface soils from around the world (Table 2-5)[20]. Of considerable interest is the high probability that all of these elements have one or more functions within the plant, even though these functions have not as yet been identified. If only extremely minute traces (as parts per trillion or less) are needed, it would be nearly impossible to conduct an experiment to prove essentiality, because the plant would likely have enough at the initiation of the test to carry it through to maturity. Not only that, but the environment (air and growth medium) would likely supply the traces of the element necessary for normal plant growth.

Element	Typical Soil Content	Element	Typical Soil Content	Element	Typical Soil Content	Element	Typical Soil Content
Ag	0.03-0.09ppm	Er	2-3ppm	Mo	1-3ppm	Se	0.2-0.7ppm
Al	1-5%	Eu	1-2ppm	N	0.1-.05%	Si	15-35%
As	5-15ppm	F	100-400ppm	Na	2.0-3.5%	Sm	4-6ppm
Au	1-2ppb	Fe	0.5-5%	Nb	5-50ppm	Sn	0.9-1.4ppm
B	30-80ppm	Ga	10-20ppm	Nd	28-35ppm	Sr	100-300ppm
Ba	200-700ppm	Gd	3-5ppm	Ni	10-50ppm	Ta	0.5-3.0ppm
Be	1-4ppm	Ge	0.8-1.5ppm	P	200-1000ppm	Tb	0.6-0.7ppm
Bi	0.2-0.5ppm	Hf	2-20ppm	Pb	15-50ppm	Te	1-10ppm
Br	10-50ppm	Hg	0.1-0.4ppm	Pd	50-150ppb	Th	4-10ppm
C	0.5-2.5%*	Ho	0.4-0.6ppm	Po	10-100Bq/kg	Ti	200-800ppm
Ca	0.5-5%	I	1-10ppm	Pr	7-8ppm	Tl	0.1-1.0ppm
Cd	0.2-1.0ppm	In	0.1-0.5ppm	Pt	20-60ppb	Tm	0.2-0.6ppm
Ce	30-50ppm	K	1-3%	Ra	1ng/kg	U	1-5ppm
Cl	100-300ppm	La	30-40ppm	Rb	30-100ppm	V	60-200ppm
Co	2-12ppm	Li	20-80ppm	Re	1-3ppb	Y	10-100ppm
Cr	20-100ppm	Lu	0.3-0.4ppm	S	300-600ppm	Yb	2-3ppm
Cs	1-15ppm	Mg	0.1-1.0%	Sb	0.6-1.5ppm	Zn	20-150ppm
Cu	15-35ppm	Mn	300-800ppm	Sc	5-12ppm	Zr	70-300ppm
Dy	4-5ppm						

Table 2-5

*Mineral soils

Note: Some elements are not included due to lack of adequate data.

Nutrient Release and Availability

While the figures in Table 2-5 show the range of *total* amounts of elements in the soil, they say little about the *plant-available* levels of particular elements. For example, potassium can be present at high levels in the soil as feldspars and micas, but only a tiny fraction will be extractable by plants at any one time (Figure 2-10)[21]. Fixed potassium is released slowly primarily by the weathering of the edges of mica and clay lattices (such as illite), wherein the potassium trapped between successive layers is able to escape (Figure 2-11). Microorganism activity and the release of organic acids speeds weathering of these mineral lattices, helping to make potassium available more quickly.

Nitrogen is primarily stored (immobilized) within the organic matter fraction of the soil and is released (mineralized) slowly during its decomposition (Figure 2-12). A 5% organic matter content of the soil — about 100,000 pounds

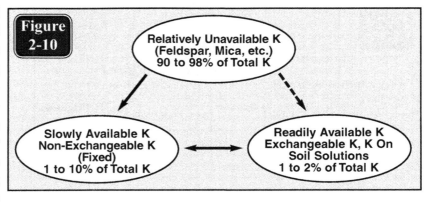

Figure 2-10

Relatively Unavailable K (Feldspar, Mica, etc.) 90 to 98% of Total K

Slowly Available K Non-Exchangeable K (Fixed) 1 to 10% of Total K

Readily Available K Exchangeable K, K On Soil Solutions 1 to 2% of Total K

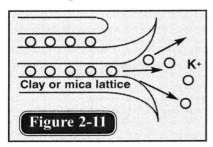

Clay or mica lattice

K+

Figure 2-11

— will contain about 5% nitrogen, or around 5,000 pounds. Only a small portion of this will be made available to plants by microbial activity during any given growing season. Of course, there are other sources of nitrogen production besides release from organic matter, such as fixation by symbiotic and nonsymbiotic organisms, rainwater additions, and fertilizer applications .

For several elements, such as phosphorus, boron, and certain of the micronutrients like zinc and copper, much of the storage may be in organic matter. Their release is mediated by microbial activity, as for nitrogen, phosphorus, potassium, and virtually all of the

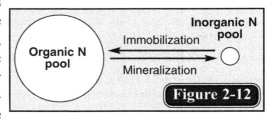

Organic N pool

Inorganic N pool

Immobilization

Mineralization

Figure 2-12

elements to one degree or another. Table 2-6 shows the percentage of total soil phosphorus in the organic fraction[22].

Iowa soils	Total soil P, ppm	Percent in the organic fraction	
Prairie soils	613	41.6%	Table 2-6
Gray-brown podzols	574	37.3%	
Planosols	495	52.7%	

The majority of phosphorus for most plants has been discovered to be taken up from the soil and translated to roots by mycorrhizal fungi. These fungi extract phosphorus from both organic and inorganic sources. Phosphorus is normally very slowly or slowly available; perhaps only 1% of the phosphorus pool is available at any one time. Note the diagram in Figure 2-13[23].

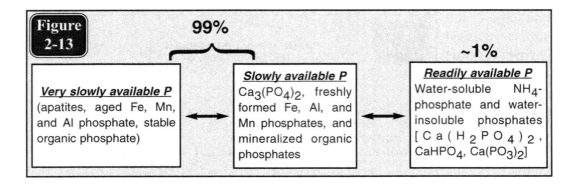

Nutrient Essentiality and Functions

Nearly all of the many elements of the Periodic Table may eventually be found to have some function with plants, but as of today only 17 are recognized as essential for carrying a plant through its vegetative and reproductive stages. These are as follows:

C HOPKINS CaFe Mg B Mn CuZn Cl Mo Co

["See Hopkins Cafe managed by mine cousin Clarence Mo and company".]

Recently, it has been discovered that nickel (Ni) may be necessary for some plants. Other elements not essential to plants *are* essential — or at least highly beneficial — for human and animal health, and normally are taken up in adequate amounts to supply the body's needs. These elements so far recognized are iodine (I), selenium (S), silica (Si), chromium (Cr), sodium (Na), and fluorine (F). Likely future candidates for beneficial elements for human and animal health include lithium (Li), vanadium (V), and germanium (Ge). Because such minute amounts of others may be needed, it may be impossible to ascertain their necessity for health because they are absorbed even from the air we breath.

At this point it is wise to list some functions of the various essential plant nutrients within the plant.

C (carbon) — the backbone of all organic compounds within the plant, be they carbohydrates, proteins, lipids, or any of other assorted compounds.

H (hydrogen) — protons, or the common element that surrounds carbon, nitrogen, sulfur, and other elements in various compounds.

O (oxygen) — the major constituent of many compounds, especially those containing acidic (-COOH) and hydroxyl (-OH) groups, like sugars and organic acids.

K (potassium) — essential for starch formation and the translocation of sugars; necessary for chlorophyll formation; important in grain formation in cereal crops, giving plump, heavy kernels; essential for tuber and root development;

promotes strong roots and stalks to prevent lodging; increases disease resistance; balances effects of both N and P[24].

P (phosphorus) — enhances cell division and fat and albumin formation; promotes flowering, fruiting, and seed formation; helps mature crops, countering the effects of excess N; promotes root development, especially lateral and fibrous roots; increases straw strength in cereal crops, helping prevent lodging; improves crop quality, especially in forages and vegetables; enhances disease resistance[25].

N (nitrogen) — encourages vegetative growth; increases grain plumpness and protein content in cereals; governs to a large degree the utilization of K, P, and other elements; produces succulence in plants, desirable for leafy crops like lettuce and radishes; a major constituent of proteins (and their enzymes), nucleic acids, vitamins, alkaloids, and many other compounds[26].

S (sulfur) — a constituent of amino acids (methionine and cystine); a component of the vitamins biotin and thiamine; determination of protein structure through S-containing functional groups; activation of some enzymes due to S-S linkages; essential in plant metabolism[27].

Ca (calcium) — effective in reducing disease incidence; aids in strong root, stem, and leaf development; improves the efficiency of photosynthesis; forms the structural components of cell walls and other plant parts[28].

Mg (magnesium) — a constituent of chlorophyll, and thus critical for photosynthesis; important in phosphorus metabolism, and interacts with several other elements; an activator of several enzymes[29].

B (boron) — essential for water absorption; necessary for the translocation of sugars[30].

Mn (manganese) — a metallic activator of certain enzymes involved in plant metabolism; essential for certain nitrogen transformations[31].

Cu (copper) — an essential electron carrier in certain oxidation-reduction enzyme systems; important in cellular respiration; necessary for iron utilization[32].

Fe (iron) — important as an enzyme activator and electron carrier in oxidation-reduction reactions of certain enzyme systems; essential in chlorophyll formation and protein synthesis in chloroplasts[33].

Zn (zinc) — necessary as a metallic enzyme activator in plant metabolism; important in the formation of some growth hormones; essential in the reproductive process[34].

Cl (chlorine) — involved in both root and top growth[35].

Mo (molybdenum) — an essential electron carrier in oxidation-reduction enzyme systems; necessary for certain nitrogen transformations; highly important for both symbiotic and nonsymbiotic nitrogen fixation[36].

Co (cobalt) — an essential element in symbiotic nitrogen fixation (a component of Vitamin B_{12}, necessary to form hemoglobin in N-fixing nodule tissue)[37].

Nutrient Interactions

Many elements interact with one another within the plant for the benefit or detriment of overall growth. Some of the major interactions include the following in Table 2-7[38].

Antagonistic	Synergistic
Cu x Fe	N x micronutrients
Fe x Mn	Mg x P
Zn x Fe	P x Mo
K x B	Si x P
Ca x B	Table 2-7

Many other nutrient interactions occur within plants and soils. Some of these are summarized in the bibliography[39]. These complimentary effects of the elements show that they serve one another if they are reasonably balanced in the soil. On the other hand, if the levels of nutrients are not balanced, they can easily elicit harmful effects on plants: disease proliferation, disrupted metabolism, and retarded growth and yields. As stated by R. Charles in a Master of Science study on food mineral content, "It is the cooperative interrelationship between minerals that fosters good metabolism"[40]. This symbiosis carries over from the plant world into the world of human and animal health as well. The use of phosphorus fertilizers changes the ratio of Ca and other elements to P in the plants, and this high P does not allow the Ca to offset potential bone decay and osteoporosis. Charles discovered in his review of research that a study of conventionally grown potatoes, as well as organically grown corn, negatively impacted the Ca:P ratio, a serious imbalance that can affect bone health[41].

Nitrogen oversupply can also cause serious problems, while a moderate supply is essential for the synergistic action of nutrients in plants and soils. Nitrogen in excessive amounts oftentimes disturbs the amino acid balance of proteins, lowering essential amino acids like lysine and methionine, while increasing nonessential ones. High nitrogen also tends to lower Vitamin C levels. This alteration in protein quality and vitamin content has powerful implications for human health[42].

Clay Minerals: Keys to Soil Fertility

Clay minerals, along with organic matter to be mentioned later, hold the key to most of the reactivity of soils. Soil chemistry involves mostly surface chemistry ... and clay minerals hold by far the greatest surface area of the soil separates. Notice the vast differences in surface areas for the separates in Table 2-8[43].

Clay contains a surface area about eight million times greater than does

	Diameter (microns)	Number of particles	Surface area (cm^2)
Clay	less than 2	> 90 billion	8 million
Silt	2 to 50	5.5 million	454
Coarse sand	500-2,000	90	11

Table 2-8

coarse sand, and nearly as much more as does silt as well. A lump of clay weighing one pound can have as much total surface area as 50 football fields[44]! It is very important to understand a bit about the nature of these clays to understand how soils work.

Clays are thought to form from the alteration of primary minerals such as micas (muscovite and biotite), and also from the decomposition of minerals such as feldspars and hornblende and then the recrystallization of the component parts of these minerals. The net result is layered clay minerals composed of octahedral and tetrahedral layers[45]. One type of clay is kaolinite, shown in the electron micrograph in Figure 2-14[46]. The somewhat conjectural scheme of formation of these multilayered clay minerals is shown in Figure 2-15[47]. The structure of these tetrahedral and octahedral layers is depicted

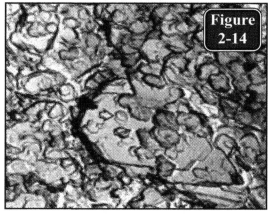

Figure 2-14

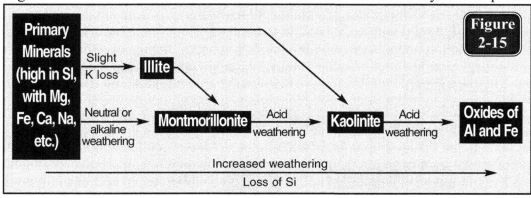

Figure 2-15

in Figure 2-16[48].

Ideally, the lattices of clay should have no charge, but in reality they have a *negative charge* on their surfaces. The charge is derived from *isomorphous substitution* and *exposed crystal edges*. The charge origin is derived as follows.

Isomorphous substitution. This occurs in both tetrahedral and octahedral layers. An element with a similar atomic diameter, but a reduced positive charge, is substituted for either Al or Si during the formation of the clay, resulting in a

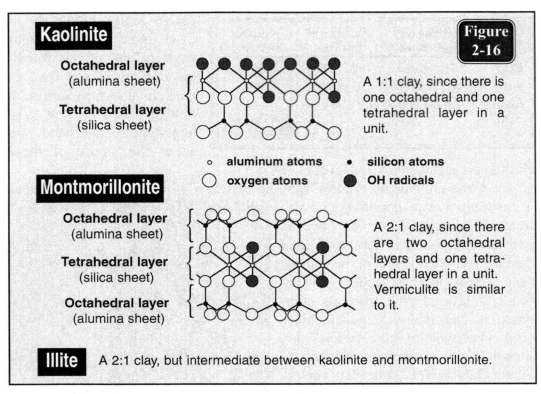

Kaolinite

Octahedral layer
(alumina sheet)

Tetrahedral layer
(silica sheet)

A 1:1 clay, since there is one octahedral and one tetrahedral layer in a unit.

o aluminum atoms   • silicon atoms
◯ oxygen atoms   ⬤ OH radicals

Montmorillonite

Octahedral layer
(alumina sheet)

Tetrahedral layer
(silica sheet)

Octahedral layer
(alumina sheet)

A 2:1 clay, since there are two octahedral layers and one tetrahedral layer in a unit. Vermiculite is similar to it.

Illite A 2:1 clay, but intermediate between kaolinite and montmorillonite.

Figure 2-16

net negative charge.

Al^{+3} replaced by Mg^{+2} = net negative charge

Si^{+4} replaced by Al^{+3} = net negative charge

Exposed crystal edges. During the process of mineral weathering or clay genesis reactions, the edges of the clay lattices may break, exposing hydroxyl

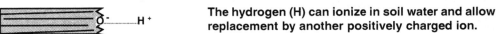

The hydrogen (H) can ionize in soil water and allow replacement by another positively charged ion.

groups that can act as ion exchangers. The hydrogen ion can exchange with other cations such as Ca^{+2}, Mg^{+2}, or K^+.

When layers of dry clay minerals are stacked one atop another, some will expand when water is added, while others will not. The 1:1 type, like kaolinite (Figure 2-17)[49], has a fixed, nonexpanding distance between successive clay units, so it cannot absorb additional ions. The 2:1 type, like montmorillonite (Figure 2-18)[50], on the other hand, will absorb water and swell and hold

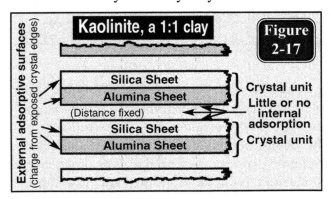

Kaolinite, a 1:1 clay

External adsorptive surfaces (charge from exposed crystal edges)

Silica Sheet
Alumina Sheet
(Distance fixed)
Silica Sheet
Alumina Sheet

Crystal unit

Little or no internal adsorption

Crystal unit

Figure 2-17

ions within the interlayers. Because of this, it can hold 10 to 12 times the number of cations (positively charged ions) that kaolinite can. Notice in Figure 2-19[51] how such a swelling clay will contract again when dried.

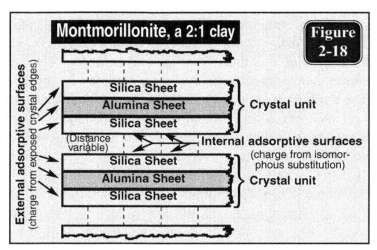

Montmorillonite, a 2:1 clay

Figure 2-18

External adsorptive surfaces (charge from exposed crystal edges)

Silica Sheet
Alumina Sheet
Silica Sheet
} Crystal unit

(Distance variable)

Internal adsorptive surfaces (charge from isomorphous substitution)

Silica Sheet
Alumina Sheet
Silica Sheet
} Crystal unit

Figure 2-19

Cation Exchange in Soils

The added charge of the isomorphous substitution and the exposed crystal edges gives rise to *cation exchange capacity*, the ability of the soil to hold cations (positively charged ions). Organic matter also contributes greatly to this capacity, in fact more than does the clay on a per weight basis. These cations include the ions of Ca, Mg, K, H, Na, and others in minor amounts. Since there is no unbalanced charge in nature for very long, the positively charged cations will latch onto the negatively charged clay interlayers and exposed edges (Figure 2-20)[52]. Organic matter contains *functional groups* that will ionize and leave negative charges.

Because these cations are held on the clay by electrostatic charges, the ions already on the lattice or organic matter can be replaced by other ions, depending

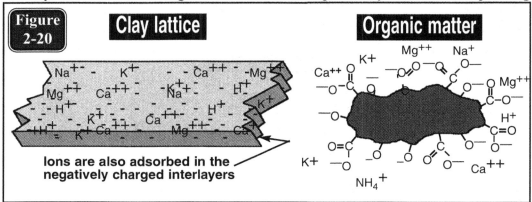

Figure 2-20

Clay lattice

Organic matter

Ions are also adsorbed in the negatively charged interlayers on the concentration of the various ions in the soil water adjacent to the surface. Then, a "mass action" reaction occurs, with resultant replacement of some attached ions by others from the soil solution. The resulting percentage of each cation that resided on the exchange complex is called its *percent base saturation*. The total of all cation percent base saturations equals 100%. Percent base saturation is expressed in terms of milliequivalents per 100 grams of soil, a milliequivalent being 1/1000 of the equivalent weight. The equivalent weight is the atomic weight divided by its combining power in chemical reactions (called its "valence"); i.e., the potassium cation (K^+) will unite with one negative charge, so its atomic weight (39 grams / mole) divided by its valence of one equals 39 grams ... or 0.039 grams for its milliequivalent weight. Calcium, on the other hand, weighs 40 grams/mole, and has a valence of two; it will neutralize two negative charges. Thus, the equivalent weight of calcium is 20 grams, and the milliequivalent weight is 0.020 grams.

The order of binding strength of the ions to the colloid is Ca>Mg>K>Na>H[53]. Due to this fact, many soils have a relative base saturation of about 50 to 80% Ca, 10 to 30% Mg, 2 to 8% K, 0 to 5% Na, and H making up the balance. Hydrogen appears on the exchange complex only in acid soils: the more acidic, the more H^+ that will attach to the exchange sites since acid soils, by definition, have a high level of H^+ ions. Highly acidic soils — pH 4 to 5 — may contain 20 to 30% H^+ on the exchange complex, replacing nutrient cations so that the soil is relatively infertile until amended by plant nutrient ions. Soils become acidic in humid climates because, over many years, a reasonably high rainfall level encourages a prolific growth of vegetation. The roots release H^+ ions in exchange for nutrient ions, thus gradually depleting the Ca, Mg, and K on the clay and organic matter and replacing it with H. Decaying organic matter also adds its share of acids that replace nutrient ions.

Mass action occurs conceptually as shown in Figure 2-21.

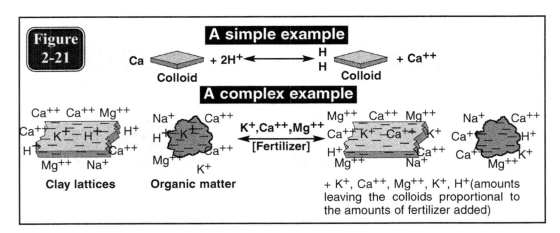

The clays affect much more that cation exchange capacity in soils. They also affect other physical soil properties; see Figure 2-22[54].

Much more could be said about clay minerals besides their contribution to cation exchange, the soil mineral reservoir, and soil physical and chemical properties. Studies in

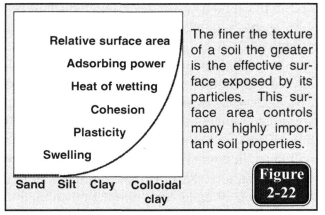

Relative surface area
Adsorbing power
Heat of wetting
Cohesion
Plasticity
Swelling

Sand Silt Clay Colloidal clay

The finer the texture of a soil the greater is the effective surface exposed by its particles. This surface area controls many highly important soil properties.

Figure 2-22

the 1980's uncovered the fact that clays can store energy and information for perhaps thousands of years. They act much like semiconductors — wild, undomesticated ones — being silicon-based and, like new semiconductors, arranged in stacks of essentially two-dimensional sheets[55]. These layers can build into many shapes, resembling piles of jewels or weird, weedy gardens.

The disorder created by isomorphic substitutions within the clay lattice serves as focal points for "amorphous domains" — domains of disorder — where catalysis occurs[56]. Compounds bind to these sites and reactions are catalyzed. Clay was the original catalyst used to refine oil. Very small amounts can accelerate chemical processes by 10,000 times or more[57]. Yet, the details of how this catalysis occurs is still obscure, prompting one Massachusetts Institute of Technology professor to state, "Our understanding of clays is worse than our understanding of biology"[58].

Clay, when wetted in water or other organic solvents, ground up, fractured, or irradiated, will emanate ultraviolet and other light wavelengths, sometimes for years[59]. Hitting a lump with a hammer will trigger a cascade of energy emissions that is maintained for months.

Moreover, clay may even replicate its complex octahedral and tetrahedral layers much like DNA does within the cell's nucleus. Such speculation on clay's formative potentials has given rise within some circles of molecular biology to the theory that clay may somehow have initiated the first "living" molecules in nature … in some ways adding meaning to Genesis 2:7, when the Lord God formed man out of the *dust* [Hebrew *aphar*, or dust, powder, clay, or mud][60] of the ground. The theory that mere synthesis of the compounds of life through clay's template — montmorillonite to DNA, for example — might have sparked the first life forms in a primordial soup, as improbable as that possibility is, cannot, of course, explain breathing the "breath of life" into the nostrils of the man to initiate actual life activity. Yet, the power of these platelets and other microscopic molecular arrangements to capture and store energy and then reemit it under certain conditions, and even replicate the molecules, is quite phenomenal indeed. These facts may explain why many ground rock powders can be very effective in stimulating plant growth[61].

Rocks, rock powders, and clays also possess magnetic qualities that have been demonstrated to be important determinants of crop growth. This quality is called *paramagnetism*, or the ability of net orbital or spin magnetic moments of the atoms and molecules of a substance to align in the direction of an applied magnetic field. Thus, in the vicinity of a magnet, the atoms and molecules of, say, basalt, will align their atomic orbitals to the field and exhibit a weak attraction to the magnet[62]. Once the magnet is removed these alignments are released, and they spring back to their original state. On the other hand, organic matter is *diamagnetic*: the atomic and molecular spins have no tendency to align with an applied magnetic field, so there is no attraction[63].

Gases also display paramagnetic or diamagnetic effects. Oxygen (O_2) is very highly paramagnetic, while Nitrogen (N_2) is diamagnetic. Water is very diamagnetic. A list of some values is given in Table 2-9[64].

Table 2-9	**Magnetic Susceptibilities of Various Substances**		

Metals

		gauss x 10⁻⁶
Fe, Co, Ni	Magnetic	—
Al	Paramagnetic	+16.5
Cu	Diamagnetic	-5.4
Zn	Diamagnetic	-11.9
Ag	Diamagnetic	-19.5
Au	Diamagnetic	-28.0
Bi	Diamagnetic	-280.1

Gases

		gauss x 10⁻⁶
O_2	Paramagnetic	+3,449.0
He	Diamagnetic	-1.8
H_2	Diamagnetic	-3.9
N_2	Diamagnetic	-12.0
CO_2	Diamagnetic	-21.0
Cl_2	Diamagnetic	-40.5
Br_2	Diamagnetic	-56.4

Nonmetals

		gauss x 10⁻⁶
Ca	Paramagnetic	+40.0
K	Paramagnetic	+20.0
Na	Paramagnetic	+16.0
Se	Diamagnetic	-25.0
Si	Diamagnetic	-3.9
S (form g)	Paramagnetic	+600.0
H_2O	Diamagnetic	-13.1

Rocks and Minerals (Kansas)

		gauss x 10⁻⁶
Shale	Paramagnetic	+18
Sandstone	Paramagnetic	+13
Bentonite	Paramagnetic	+11
Sand/gravel	Paramagnetic	+2
Oil shale	Diamagnetic	-0.003
Salt	Diamagnetic	-0.001
Limestone	Diamagnetic	-0.001

Some experiments have shown that plants respond not simply to the elements within a particular rock or mineral, but also to some sort of energetic effect from the material[65]. This is to say that one source of paramagnetic limestone, equal in composition and fineness of grind to another limestone, produces a greater crop yield response. This effect may be related to "energy" — however we might define that elusive term — which is stored within the molecular structure of the rock much like energy has been discovered to be stored in clays. These stored energies presumably are responsible for plant growth enhancement, especially near sharp corners of paramagnetic rocks. A uniform powder application on the soil would diffuse this corner effect. Until "energy" and its effects on plant growth are more thoroughly understood, it will be difficult to explain the nature of para-magnetic effects. So diverse and all-encompassing are the many intricacies of God's creation.

Soil Organic Matter

The organic matter of the soil is a most critical component, since its effects are so pervasive. It is rather humiliating to admit that this soil component is a product of death, not life. Yet, even this fact confirms the link of service within the natural world: living plants, animals, and microbes die, and their remains are recycled through the crucible of organic matter to generate life anew. This hum-

ble portion of the soil, while comprising from less than 1% up to 7 or 8% in mineral soils, but over 50% in many organic soils, carries far more influence than its portion implies[66]. Figure 2-23 shows an "average" organic matter content of 5%. It is also within this fraction that the all-important micro and macroorganisms thrive.

Soil organic matter is the accumulated residue of partially decayed and partially

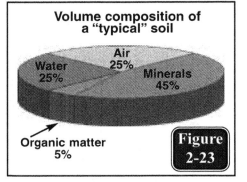

Volume composition of a "typical" soil

Water 25%
Air 25%
Minerals 45%
Organic matter 5%

Figure 2-23

resynthesized plant and animal residues. These residues may consist of plant roots, leaves and stems, animal manures or carcasses, or microorganisms that are actively attacking and decomposing these organic materials. It is thus a transitory material which must be renewed continually.

The Importance of Soil Organic Matter

Among the many claims to fame of organic matter is the fact that, for most soils, over half of the cation-exchange capacity is derived from exchange sites on the organic matter. While a montmorillonite clay may contribute perhaps 200 meq/100 grams of soil, organic matter may grant 500 meq/100 grams of soil. Organic matter also is the single most important factor in the stabilization of soil aggregates, which is discussed later.

Some of the most important contributions of soil organic matter are the following:

1. Coarse organic matter on the surface reduces the impact of raindrops and allows clear water to seep into the soil. It also reduces wind erosion.

2. Decomposing organic matter produces polysaccharides and "glues" that produce a stable, strong soil structure for rapid air and water movement.

3. Live roots decay and provide channels for further root growth and air and water movement.

4. Fresh organic matter provides food for earthworms, microbes, and other soil creatures that increase pore space and release nutrients.

5. Surface mulches moderate soil temperature like a blanket. A straw mulch can lower the temperature at 0.5 inch by 10° F, while a clear plastic mulch may increase it by 10° F.

6. Evaporation losses are reduced by surface residues.

7. Nutrients are supplied by decomposition, especially nitrogen (15 to 110 lb/acre), and cations are held by its high cation exchange capacity.

8. Fresh organic matter makes soil phosphorus more available in acid soils.

9. Organic matter buffers the soil against rapid chemical changes.

10. Some plant diseases can be controlled by organic matter additions.

11. Organic matter provides a buffer against rapid chemical changes, such as when lime and fertilizers are added.

12. Available water for growth is increased, since organic matter acts as a sponge.

Components of Organic Matter

The compounds found in the soil organic fraction are highly diverse, as might be expected, produced from a greater or lesser state of degradation from original plant and animal tissue. One means to categorize these substances is shown in Figure 2-24[67].

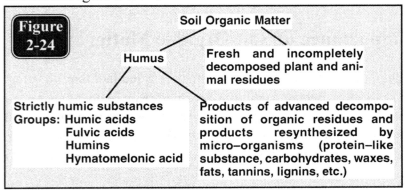

This rather old classification is based on color and behavior towards solvents. Humic acids are those dark-colored compounds extracted by alkaline solutions and precipitated during acidification to a pH of about 1.0. This group contains aromatic components with humic acid cores, plus proteins, peptides, amino acids, and polysaccharides. Fulvic acid is soluble in alkaline solutions as well as water and alcohols, while humic acid is insoluble in alkali but soluble in acids.

Individual groupings of organic compounds within organic matter include the following[68]:

1. Fats, oils, and waxes
2. Carbohydrates, including sugars, starches, and cellulose
3. Proteins and their derivatives, including amino acids, amides, purines, pyridines, pyrimidines, and albumins
4. Lignins and their derivatives
5. Tannic substances, both simple and condensed
6. Resins and terpenes
7. Pigments and derivatives, including chlorophyll, cyanins, and others
8. Minerals
9. Others: organic acids, aromatic compounds, hydrocarbons, alcohols, and others

Polysaccharides as gels and slimes, secreted by bacteria and fungi in the rhizosphere, constitute about 15 to 20% of soil organic matter which are so critical to the formation of microscopic soil aggregates that form the first tier of structural components[69]. As will be discussed later, these tiny units are loosely bound together by fungal and actinomycete hyphae to form crumbs. About 80% of the organic matter is humic material made up of complex stable polymers of carbohydrates, polyphenols, and amino acids. They consist of fibers or globular deposits about 10 mm in diameter[70]. For one view of soil humic substances, note the micrograph (105,000 magnification) in Figure 2-25[71].

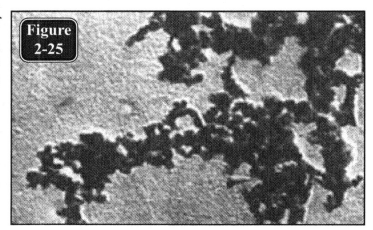
Figure 2-25

Organisms in Soil Organic Matter

The center of the organic fraction's activity lies not only in its nutrient-supplying power, but in its organism content. The number and variety of species found in soils is extremely variable, dependent on the climate, mineralogy, management methods, and type and amount of organic materials. Estimates of these numbers are also highly variable. The numbers and weights have been estimated by some researchers as shown in Table 2-10[72].

Table 2-10 Organism	Estimated number	Average weight per upper foot of soil	
		Low	High
	number / gram	lb/acre	
Bacteria	3,000,000 to 1,000,000,000	500 to 1,000	
Actinomycetes	1,000,000 to 20,000,000	800 to 1,500	
Fungi	5,000 to 15,000,000	1,500 to 2,000	
Yeasts	1,000 to 100,000	————	
Algae	1,000 to 500,000	200 to 300	
Protozoa	1,000 to 500,000	200 to 400	
Nematodes	50 to 500	25 to 50	
Earthworms	13,000 to 1,000,000	15 to 1,100	

Groups of soil organisms are highly diverse, but their appearances can be generalized. Note the wide range of types in the soil, especially within the root zone, shown in Figure 2-26[73].

Each organism, whether in the plant or animal kingdom, has its special niche within which it performs its functions. Photosynthetic algae and cyanobacteria multiply near the soil surface to intercept light. Bacteria and fungi populate all areas where carbon residues are accessible, especially along the roots where energy is exuded in abundance. Some autotrophic bacteria, such as nitrifiers and sulfur oxidizers, utilize CO_2 rather than organic carbon for their carbon source and obtain their energy largely from the oxidation of organic substances. Earthworms traverse the soil mass in search of organic materials wherever they may be found. Mites search amongst macropores for nematodes and other small creatures, while protozoa seek out bacteria and organic bits. Nematodes of all sorts graze on fungi, bacteria, plant roots, or other nematodes. Actinomycetes, like fungi, consume organic remains. Rotifers, springtails, ants, tardigrades, spiders, millipedes, centipedes, and other creatures all seek out niches in which to grow and reproduce.

It is interesting that some crops — mostly row crops and vegetables — prefer a predominance of bacteria in the soil, while woody perennials and trees prefer fungi. In soil environments receiving few or no inorganic chemical imports, the ratio of fungi to bacteria in the bulk soil is as shown in Table 2-11[74].

The reproductive potential of bacteria is enormous. If a single bacteria and its progeny divided every hour, in 24 hours the original cell would have yielded 17,000,000 cells. In six days, if allowed to multiply unrestricted, the organisms

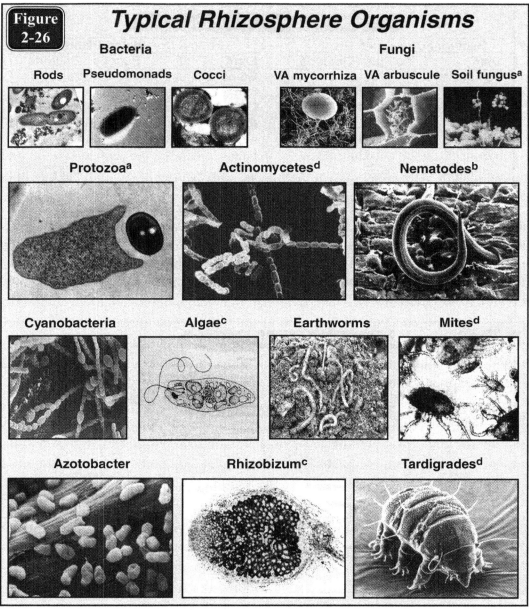

Figure 2-26

Typical Rhizosphere Organisms

Bacteria

Rods   Pseudomonads   Cocci

Fungi

VA mycorrhiza   VA arbuscule   Soil fungus[a]

Protozoa[a]   Actinomycetes[d]   Nematodes[b]

Cyanobacteria   Algae[c]   Earthworms   Mites[d]

Azotobacter   Rhizobizum[c]   Tardigrades[d]

Picture credits:[a]Kilham, *Soil Ecology*, ©1994. Reprinted by permission of Cambridge University Press; [b]Dropkin, *Introduction to Plant Nematology*, ©1989 Wiley Interscience. Reprinted by permission of John Wiley&Sons, Inc.; [c]Sylvia, et al., *Principles of Soil Microbiology*, ©1998. Reprinted by permission of Pearson Education, Inc., Upper Saddle River, NJ.; [d]Dindal, *Soil Biology Guide*, ©1990 Wiley Interscience. Reprinted by permission of John Wiley & Sons, Inc.

would surpass the earth in volume[75]!

Production of Humic Substances

The pool of soil organisms is continually in flux, some actively growing

Fungal – Bacteria Ratio for Crops			
Bacterial Dominated Plants		**Fungal Dominated Plants**	
Plant	Fungus : Bacteria	Plant	Fungus : Bacteria
Turf grass	0.9 to 1.5	Grape	3 to 5
Broccoli	0.3 to 0.7	Deciduous trees	10 to 100
Kale	0.5 to 0.8	Coniferous trees	100 to 1000
Carrots	0.8 to 1.5	Alder	5 to 100
Corn	0.8 to 1.0	Apple	10 to 50
Wheat	0.8 to 1.2	Pine	50 to 100
Lettuce	0.8 to 1.0		
Tomato	0.8 to 1.0		**Table 2-11**
Tobacco	1.0 to 3.0		

and dividing, while others die and return to the inactive organic phase. By-products of the organisms which have fed on organic residues, and their decomposition products once they have died, produce the very complex mileau of humic acids, fulvic acids, humin, and other components of soil organic matter. The process may be viewed in Figure 2-27[76].

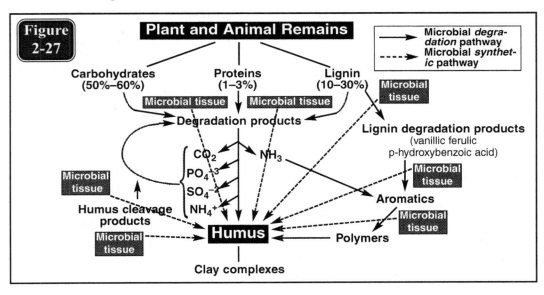

Humic substances are readily able to couple with clay minerals due to cationic and anionic exchange reactions, as well as hydrogen bonding and Van der Waals forces. One such scheme of complexing is shown in Figure 2-28[77].

Organic carbon compounds can persist in soils for long periods of time under the right conditions, although there is a constant turnover of fresh materials into this persistent fraction. Using carbon-14 (^{14}C) dating methods, it is possible to determine the average age of modern humus to obtain a *mean residence time*. For fractional portions of organic matter from Chernozem soils, the average age of

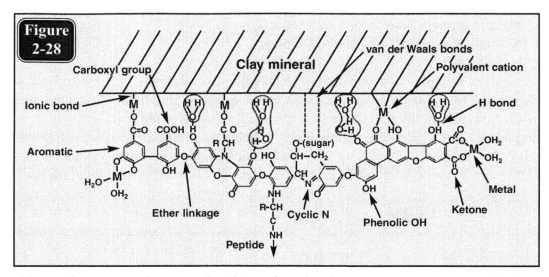

Figure 2-28

this persistent organic matter has been found to be 25 to 1,400 years[78]. The total humic acid portion of this sample averaged 1,235 years; a podzolic forest soil had organic matter which was less stable. Buried soils were found to have organic matter dating back 20,500 years[79].

Soil Structure and Bulk Density

A major function of soil organic matter is to help create and stabilize strong structural units, sometimes called "peds". *Structure* is the gross aggregation or three-dimensional arrangement of soil particles. These structural units can vary from *single grains* (as sand) to *massive* (solid, no cleavage planes). A soil may have different types of structure at different layers of the profile. Several of the types of aggregates commonly found in soils are shown in Figure 2-29[80].

Structural units are formed by complex, poorly understood processes. Among the more important factors involved are the following[81]:

1. Wetting and drying
2. Freezing and thawing
3. Physical action of roots
4. Organic decay and soil

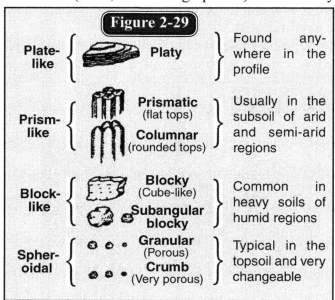

Figure 2-29

Plate-like	Platy	Found anywhere in the profile
Prism-like	Prismatic (flat tops)	Usually in the subsoil of arid and semi-arid regions
	Columnar (rounded tops)	
Block-like	Blocky (Cube-like)	Common in heavy soils of humid regions
	Subangular blocky	
Spher-oidal	Granular (Porous)	Typical in the topsoil and very changeable
	Crumb (Very porous)	

organisms, their activity and by-products

5. Modifying effects of adsorbed cations

6. Soil tillage

The interplay of these factors creates lines of weakness, forcing soil particles to and fro, causing contact of one to another and gluing them into aggregates of various configurations. One may visualize the swelling and contracting of wetting and drying, or freezing and thawing, to help create cleavage planes (Figure 2-30)[82]. Roots form channels and force soil particles apart. Earthworms create multitudes of channels throughout the soil mass, and their castings are fairly stable against breakdown; they vary according to the organic residues available which serve as food (if the earthworms have not been poisoned out of existence), and may vary from 13,000/acre (no manure added) to 1,000,000/acre (manure added)[83]. They may leave up to 16,000 lb/acre of castings each year!

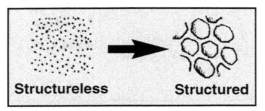

Structureless **Structured**

Figure 2-30

Soil organisms such as bacteria, fungi, algae, cyanobacteria, and others break down organic residues and produce various by-products, one of which is *polysaccharide*. This general group of compounds involves strings of sugar molecules which act as a glue to bind soil particles together. As little as 0.02% of added polysaccharide in the soil will markedly improve soil structure[84].

Not to be minimized is the effect of symbiotic mycorrhizae on structure formation. These fungi extend thread-like hyphae from plant roots out into the soil in all directions, stabilizing cleavage surfaces and creating sac-like structures around particles. The issue of microbial effects on structure will be discussed in more detail later.

A predominance of sodium in the soil will tend to disperse soil particles, while calcium in abundance tends to flocculate them (cause them to clump). Tillage, such as plowing — and especially when the soil is wet — on the other hand, will weaken and destroy structural units.

Figure 2-31

Note how a low organic matter soil on the left of Figure 2-31[85] has produced a puddled, massive structure with few cleavage planes, while the well-granulated soil on the right has excellent porosity. This superb structure is typical of grassland soils that are highly productive.

Just as important as the type of structure is its strength and stability. Microbial glues are needed to bind particles together; iron oxides also help cement them. In general, the larger the aggregates present in any particular soil, the lower their stability. An electron micrograph of a soil ped is shown in Figure 2-32, with a polysaccharide core (and fungus within it), and clay platelets glued around it[86].

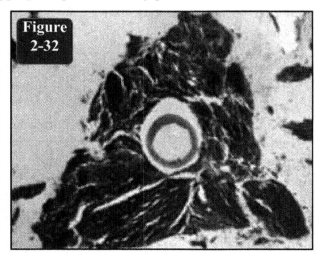

Figure 2-32

The overriding factor common to excellent, strong soil structure is *organic matter*. Aggregates are more stable when certain binding agents — such as polysaccharides and humic acids — are present in sufficient amounts. These are produced in larger quantities when a vigorous microbial population is working within a highly organic soil. Aggregates in low organic matter soil fall apart easily, while those from high organic matter soil are more stable.

Any organic matter added to a soil — compost, manure, leaves, cover crops, etc. — provides food for microbes, which in turn create the units. For example, plain sewage sludge can markedly improve the number of water stable aggregates in a soil (Figure 2-33)[87].

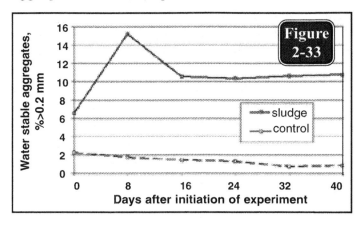

Figure 2-33

The value of a strong soil structure with cleavage planes in both vertical and horizontal directions is multi-faceted. Besides creating a higher percentage of macropores through which air, water, roots, and soil creatures can more easily move, the peds store nutrients — especially nitrogen — within the ped. This environment is quite anaerobic, which restricts the activity of bacteria and other organisms, tending to protect the nutrients from loss until a root and its highly active rhizosphere move within close proximity. Thus, both aerobic and anaerobic sites exist side-by-side in a well-structured soil. Both are desirable; both serve definite, mutualistic functions.

The mechanisms of structure formation are similar for both sandy and clayey soils, with some minor differences as indicated below.

Sandy soils. Organic matter and microbial by-products bind sand particles together to improve water-holding capacity, infiltration, aeration, porosity, cation exchange capacity, and other properties.

Clayey soils. Organic matter and microbial activity build cleavage planes for a strong, stable structure while increasing the number of macropores to improve air and water movement, root growth, and percolation and infiltration. Puddling of the soil is reduced.

Compaction of the soil at a particular layer, such as at the plow layer, essentially destroys the structural development at that plane and greatly restricts water and air movement, root penetration, and crop yields. Note bulk densities in the soil profile shown in Figure 2-34[88].

Bulk density is a measure of the weight of dry soil per volume of soil. To a point, the less

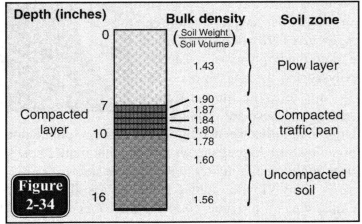

Figure 2-34

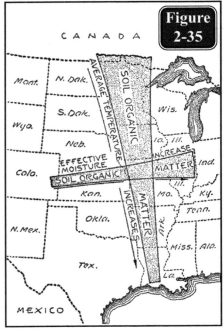

Figure 2-35

the bulk density the better, since less density means more pores, implying better structure and ability for air and water to move through the mass and for roots to grow within a highly aerobic environment. Clays and loamy soils have bulk densities of around 1.0 to 1.6 grams/cc, while sandy soils may be 1.2 to 1.8 grams/cc. Hardpans can range from 1.8 to 2.0 grams/cc or even higher. Bulk density tends to increase with soil depth, mostly because the organism, organic matter, and weathering effects that create a strong structure become less effective with depth. Also, the weight of layers above tends to compact layers below due to gravity.

Soils vary greatly in organic matter, microorganism content, bulk density, and structural integrity depending on climate, topography, parent material, time, and tillage practices. Notice the map in Figure 2-35 that illustrates how organic matter increases with increasing rainfall from west to east across the United States, and with lower average temperature from south to north[89]. Greater vegetative growth from a more humid climate leads to more residues for organic matter creation, and colder temperatures to the north arrest the mineralizing activities of microorganisms in the winter, allowing more organic materials to accumulate.

Prairie and grass ecospheres nearly always lead to higher organic matter and better structured soils than do forest ecospheres. A big reason for this difference is the immense amount of fibrous roots in the topsoil of a grassland, the decaying roots providing a high annual return of residues to the soil besides an abundance of leaves, whereas trees provide many fewer fine roots in the surface soil. Note the development of grassland and forest soils in Figure 2-36[90].

Tillage nearly always reduces soil organic matter, and thus tends to restrict aeration and water movement as soils become more compact. This leads to more runoff and erosion and less available water for crop growth,

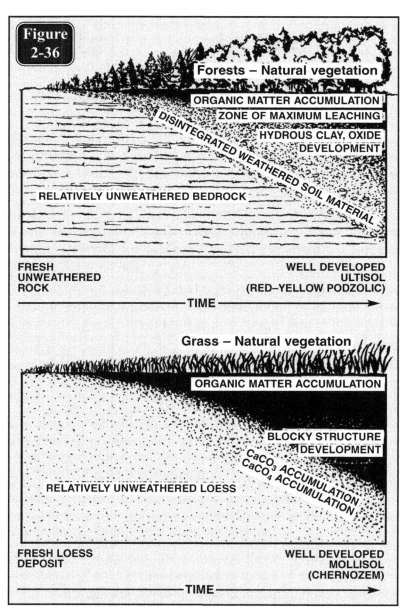

Figure 2-36

Forests – Natural vegetation
ORGANIC MATTER ACCUMULATION
ZONE OF MAXIMUM LEACHING
DISINTEGRATED WEATHERED SOIL MATERIAL
HYDROUS CLAY, OXIDE DEVELOPMENT
RELATIVELY UNWEATHERED BEDROCK

FRESH UNWEATHERED ROCK
WELL DEVELOPED ULTISOL (RED–YELLOW PODZOLIC)
TIME

Grass – Natural vegetation
ORGANIC MATTER ACCUMULATION
BLOCKY STRUCTURE DEVELOPMENT
$CaCO_3$ ACCUMULATION
$CaCO_4$ ACCUMULATION
RELATIVELY UNWEATHERED LOESS

FRESH LOESS DEPOSIT
WELL DEVELOPED MOLLISOL (CHERNOZEM)
TIME

resulting in lower yields. Notice in Figure 2-37 how the average organic matter content of three North Dakota soils, before and after an average of 43 years of cropping, drastically reduced soil organic matter. About 25% of the organic matter was lost from the 0 to 6-inch layer as a result of cropping[91].

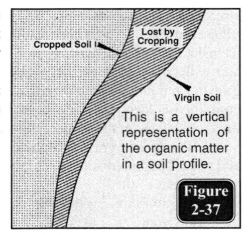

Cropped Soil
Lost by Cropping
Virgin Soil

This is a vertical representation of the organic matter in a soil profile.

Figure 2-37

Soil Air

The Gases in Soil Air

All of the gases found in the air are found in the soil. As discussed earlier, the soil contains about 50% pore space, about half of which may be occupied by water and half by gases. Within the 25% or so occupied by soil gases, the oxygen level is generally less than that of the atmosphere, but CO_2 is much more, while nitrogen is usually more as well. Different soils contain different percentages of these three major gases, depending upon the rate of decomposition of organic compounds ($[CH_2O]_x + O_2 \longrightarrow CO_2 + H_2O$), the rate of denitrification ($[NH_2]_x \longrightarrow NO_2^- \longrightarrow NO_3^- \longrightarrow N_2$), and the rate of diffusion of gases into and out of the soil. The diffusion rate is governed by the porosity, in particular the percentage of large pores (macropores) and small pores (micropores), as well as their continuity ... especially as expedited by worm and root channels. The more large pores and interstitial channels for gas movement, the greater the rate of exchange between the air above and the air below the soil surface, and the more they will resemble each other.

Notice how the Iowa bulk soil shown in Figure 2-38 has oxygen and nitrogen gas contents nearly identical to air, but the CO_2 level is nearly seven times higher than air[92]. The New York soil, which is presumably of finer texture with more micropores and more compaction, is 39% lower in oxygen with a 150-fold increase in CO_2. Nitrogen is also 2.4 percentage points higher. All of these changes point towards a slow diffusion of gases, and likely a predominately anaerobic condition, perhaps due to compaction, that would encourage denitrifying bacteria, which are anaerobic, to reduce nitrate to atmospheric dinitrogen ($NO_3 \longrightarrow N_2 + O_2$). These values are representative only of the average condition of the soil, not of the soil in the rhizosphere that is rapidly depleted of its oxygen due to intense microbial metabolism as bacteria and the other microbes feed on root exu-

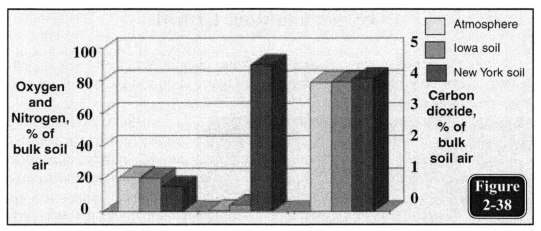

Figure 2-38

dates, rapidly depleting oxygen along the root to create anaerobic microsites. This issue will be discussed when rhizosphere functions are addressed.

The gases in soil air are essential for living processes of plant root cells and soil organisms. They conspire in a mutualistic effort to supply the metabolic cycles with the needed raw gaseous materials for life. Without sufficient oxygen a root will slow in its growth and eventually stop entirely, whereas if growing with plenty of oxygen it can extend three inches per day[93].

It would be expected that as one explores greater depths of the soil profile, the oxygen level would decrease. Such is indeed the case, especially with heavier-textured soils that allow slower gas movement. Notice Figure 2-39, which depicts the oxygen level with soil depth in a fine-textured soil having fewer macropores for gas exchange versus a coarser-textured soil having many larger pores[94].

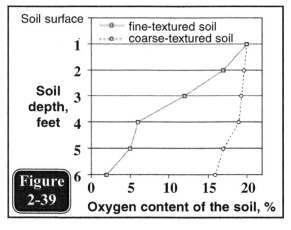

Figure 2-39

Seasonal variations in soil gases also occur in soils. High moisture conditions in rainy seasons tend to lower oxygen and raise CO_2 levels due to higher root and microbial metabolism as well as slower diffusion of gases through the soil. Higher temperatures dry soils more quickly, opening soil pores and allowing for soil air to more closely resemble the atmosphere. However, if growth of roots and microbes is especially rapid, soil CO_2 levels may actually rise and oxygen levels drop with high temperatures.

Oxygen and Root Growth

The effect of oxygen on root growth has already been discussed at some length. Note the profound effect that oxygen deprivation can have on the proliferation of roots in Figure 2-40[95].

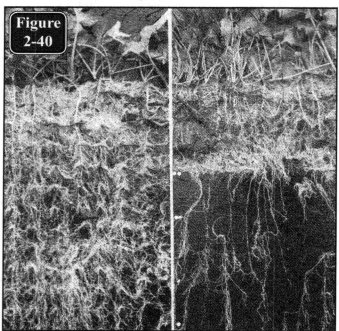

Figure 2-40

When oxygen levels are reduced in the soil, the active, energy dependent mechanisms by which nutrients are taken up are stressed. As a result, both nutrients and water uptake are reduced. The reasons for these reductions relate to less energy being available in ATP (adenosine triphosphate) and other energy-trapping compounds, which is theorized to be the primary means of moving water, minerals, and organic molecules through root cell membranes. Ion carriers can diffuse across the relatively fluid proteins in the core of the layer to carry ions to the opposite side (Figure 2-41)[96].

Carriers of some sort are involved in shuttling nutrients across the cell membrane, resulting in accumulations of ions to much higher levels inside than outside the membrane. However, this scheme does not take into account the vigorous activity of bacteria, fungi — especially mycorrhizae — actinomycetes, algae, and other microbes which ferry nutrients to the root surface and/or make them available for uptake. This marvelous synergism of soil organisms, water, oxygen, root cells, and organic and inorganic carriers within

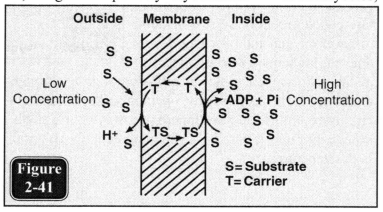

Figure 2-41

Outside   Membrane   Inside

Low Concentration   High Concentration

ADP + Pi

S = Substrate
T = Carrier

44

root cell membranes provides an incredibly complex example of cooperation amongst multiple structures and components of structures to bring nutrients into the root and then up into the plant canopy. Details of these mechanisms will be discussed later; for now note a summary of them in Figure 2-42[97].

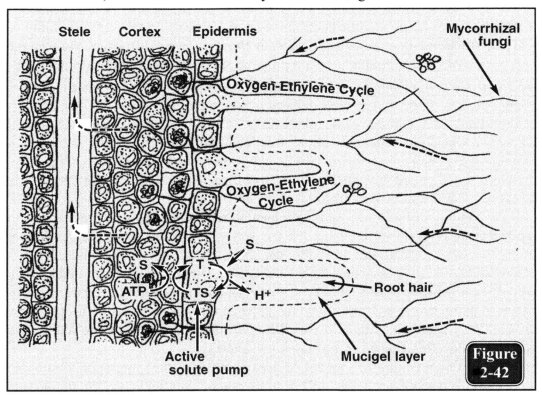

A lack of adequate oxygen in the soil due to waterlogging or compaction will also cause microbes to create toxic by-products that are highly detrimental to plant growth. Some of these toxic substances include methane from carbohydrate breakdown ($C_6H_{12}O_6 \longrightarrow 3CO_2 + 3CH_4$), hydrogen sulfide from sulfur reduction (R-SH $\longrightarrow$ H_2S), organic acids (acetic, lactic, propionic, butyric, citric, etc.), and others. Denitrification — the changing of nitrate to N_2 or nitrogen oxides that escape to the atmosphere — is another undesirable side effect of low soil oxygen levels. In addition, in low oxygen soils mineral elements are reduced to a lower oxidation state, which yields a much more soluble compound that can accumulate to toxic levels ($Fe^{+3} \longrightarrow Fe^{+2}$; $Mn^{+4} \longrightarrow Mn^{+2}$; $Cu^{+3} \longrightarrow Cu^{+2}$). Very local areas of anaerobic activity along the root surfaces caused by temporary oxygen depletion is actually highly beneficial to the plant, as will be shown later, but in this case the bacterial population remains aerobic; its activity is merely slowed down by ethylene and low oxygen levels.

Soil Temperature

The importance of soil temperature to the activity of the soil organisms and roots should be obvious. For every increase or decrease in temperature, biological activity increases or decreases, respectively, from two to three times (called the "Q_{10}" in biology), although there are pronounced differences in the ability of different plants and organisms to grow at different temperatures. For instance, *nitrifying bacteria*, those that convert reduced nitrogen forms to nitrate, do not become active until soil temperatures reach about 40°F[98]. Other organisms, called *cryophilic* species, perform well at less than 40°F, while some, called *thermophiles*, do well even above 100°F. *Mesophiles* are those that survive best between the high and low extremes. Roots of oats grow best at about 60°F, while roots of potatoes thrive best from 60 to 70°F[99].

Not only is biological activity related to temperature, but it is related to soil chemical activity as well. Thus, as the soil temperature drops, the ability of microbes to make nutrients available decreases. Both plant roots and microbes decrease their activity together, however, so they tend to match plant uptake with soil release. In spite of this fact, some nutrients, like phosphorus, become very difficult to extract from soil phosphate minerals as the soil temperature drops below 50 to 55°F. The result can be a purplish color of plant leaves due to inadequate sugar translocation caused by a phosphorus shortage.

The temperature of the soil is dependent upon the geographical site one considers, and since each site differs from others in latitude and longitude and many other factors, temperatures and temperature profiles differ. Some of the factors involved in soil temperature at a specific location are as follows:

1. Aspect and slope (angle at which the sun's rays strike the surface)
2. Season, dependent on latitude
3. Prevailing air masses and their movement relating to ...
    a. Rainfall: total, frequency, and intensity
    b. Wind speed and direction
    c. Humidity
    d. Cloudiness and solar radiation
4. Surface vegetation and ground cover
5. Tillage and surface disturbance, if any
6. Soil structure and permeability to air and water
7. Soil density, especially of a mineral versus an organic soil

In summary (Figure 2-43)[100], the heat of the soil surface is the product of energy gained versus that which is lost. Sunlight energy strikes the soil surface and some is reflected; the lighter the color, the greater the reflection. Some heat is transmitted from the air to the soil, while a greater portion is radiated into the atmosphere and space. Soil

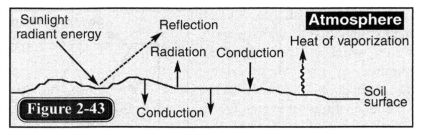

Figure 2-43

water evaporation cools the earth due to water's requirement of 585 calories to evaporate one gram at 68°F. This cooling effect of only one pound of water can lower the temperature of a cubic foot of a typical mineral soil by 28°F. The resultant heat at the soil surface can then be conducted down through the soil to lower levels, dependent on soil density and porosity. The heat actually entering the soil is greatly influenced by surface cover. Any mulch, crop, or shade cover acts as a blanket on soil temperature changes.

Diurnal fluctuations in surface soil temperature reveal that soil temperatures reach their peak an hour or two later than the afternoon temperature peak. This is due to the continued absorption of radiant energy by the soil throughout the day. A surface mulch greatly reduces both the absolute temperature and the daily variation in temperatures, as noted in the graphs of Figure 2-44[101].

Over the course of a year in areas outside the tropics, the soil surface temperature can change drastically due to seasonal fluctuation. As the average surface temperature increases in summer, the temperature front lags as heat is transferred by conduction into the subsoil. The reverse is true as the temperature decreases in winter. Below about 12 feet, the temperature varies little, since the transmission of surface fluctuations is overshadowed by the constant heat transmission of the earth at greater depths. Figure 2-45[102] encapsulates these annual soil temperature profiles over the course of a year.

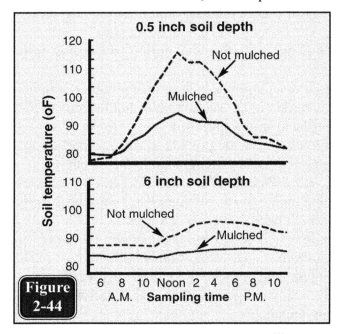

Figure 2-44

Soil Water

Properties of Water Important to Soils

Water, of course, is crucial to

the existence of all life. A cell may contain 70 to 90% or more water, which is necessary for a number of critical functions. These functions depend upon the unique qualities of water which no other compound on earth possesses.

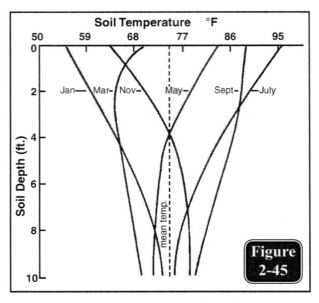

Figure 2-45

1. *Water absorbs much heat while changing in temperature*, more so than most anything else in nature. This is called **specific heat**, the input of heat energy required (in calories) to raise the temperature of any substance 1^0C[103].

Substance	Specific heat (cal/g^{0}C)
Water	1.00
Wood	0.42
Chalk	0.21
Glass	0.19
Granite	0.19
Iron	0.11
Zinc	0.09

As a result of this property, cells of plants, animals, and microbes are greatly stabilized.

2. *Water conducts heat well*, allowing heat generated during metabolism to be dissipated.

3. *Water has a very high surface tension*, greater than for any other common liquid. Because of this strong cohesion of the surface layer, liquids tend to pull into a round shape. When translated into a pulling force, pure water — if it could be made — would require over 200,000 pounds of force to pull apart a one inch diameter column, giving it a tensile strength similar to steel[104].

4. *Water is highly adhesive to other substances but will not wet fats and waxes.*

5. *Water is transparent*, so visible radiation can be transmitted through it to the interior of cells and organs in which photosynthesis can occur.

6. *Water is rather inert chemically, so it can serve as a medium for nearly all of the chemical reactions within plants.* Because it does not take part in the reactions — the exception being photosynthesis — it is in reality a *catalyst*.

7. *Water decreases in viscosity as the temperature increases*, a factor influencing reaction rates and solute movement.

8. *While pure water is a good insulator, water having dissolved ions is a good conductor for electrons.*

Other qualities of water that relate to living systems are its *high latent heat of vaporization*, or the amount of heat necessary to turn one gram of water into steam (539.3 calories/gram), and a *fairly high heat of fusion*, the amount of heat needed to change one gram of ice into water (79.7 calories/gram). Water also *expands as the temperature cools from 4^0 to 0^0C and forms ice*. This expansion is due to the polar orientation that water assumes within the crystal structure. Note this polar orientation of water molecules in Figure 2-46[105].

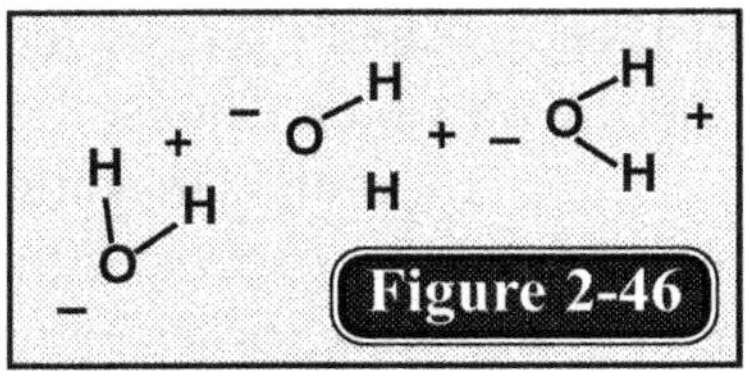

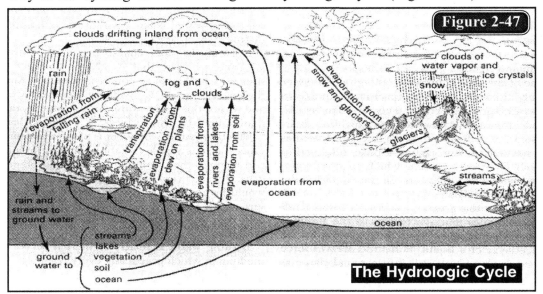

Figure 2-46

Water is indeed the uniquely created liquid that serves as the "universal medium" for life. It is analogous to the spirit of God that showers the minds and hearts of His elect and catalyzes their thoughts and actions. Water is thus a symbiotic motivator of virtually all biochemical reactions within cells, and the universal carrier and medium for the movement of solutes into the plant via the xylem vessels, and for the movement of photosynthate and synthesized compounds through phloem vessels throughout the plant.

Water Movement in Soils

The supply of water to soils through rain or snow, its percolation into the groundwater, and evaporation to complete the hydrologic cycle have been discussed earlier. The transpiration of water from plants through root uptake in the rhizosphere, followed by evaporation from leaf surfaces, contributes in no small way to the cycling of water through the hydrologic cycle (Figure 2-47)[106]. This

Figure 2-47

clouds drifting inland from ocean

clouds of water vapor and ice crystals

rain

fog and clouds

snow

evaporation from falling rain

transpiration

evaporation from dew on plants

evaporation from rivers and lakes

evaporation from soil

evaporation from snow and glaciers

glaciers

streams

evaporation from ocean

rain and streams to ground water

ocean

ground water to { streams lakes vegetation soil ocean

The Hydrologic Cycle

evaporation from leaves greatly facilitates cooling of the leaves as well, or cooling of soil surfaces on which evaporation occurs.

Because of the properties of water mentioned above, water which falls as rain or that is applied in irrigation water flows into soil surface pores through a process called **infiltration**, and then moves through the soil mass by **percolation**. When water first flows into the soil, the surface "wets up" and fills surface pores. It then gradually moves into the soil beneath, displacing air in pores and filling them with water. The soil becomes saturated.

If rainfall or irrigation cease, the added water will continue to move downward for a day or so and then stop as gravity pulls water in macropores into less moist soil whose pores can absorb more water. The soil at this point reaches **field capacity**, when no more outflow from gravity occurs. At this stage the micropores are all filled with water, and a water film exists on the inside surface of macropores, with air reentering these larger pores.

Water Availability in Plants

Plants absorb moisture in the micropores and in films around macropores, eventually leaving only a very thin layer of water clinging to mineral and organic particles which the roots and mycorrhizae cannot remove. At this stage, the **wilting point** has been reached. If the soil is allowed to dry even further with no water additions, then the **hygroscopic coefficient** is reached. These stages of soil water content are shown in Figure 2-48[107]. It will be noted that a considerable amount of water still remains clinging to soil surfaces when the wilting point has been reached, in fact as much water on the particles of soil as the plant has used after field capacity is reached. Unavailable hygroscopic water comprises a significant portion of soil water that plants can never use because it adheres to soil solids with great force. The forces holding water to soils are depicted in Figure 2-49[108].

A pressure of 15 atmospheres or more is required to remove the water on soil particles that is not extracted by plant roots. Water at the hygroscopic coefficient is removed by a force of 31 atmospheres, while water at the soil-water interface, directly adhering to particles, needs 10,000 atmospheres to remove it[109].

Water on soil particles and in pores between field capacity and the hygroscopic coefficient — most of which is available to plants — is called **capillary water**. This water can move toward the soil surface or toward root surfaces as water is removed by root uptake or surface evaporation. The movement of solutes in this capillary water, like sodium from salty irrigation water, will cause a deposition of salts near and on the surface which can be detrimental to crop growth.

The absorption and movement of water in different soils can differ dramatically as clay and organic matter increases ... and thus total surface area on the colloid. A finer textured and/or higher organic matter soil will possess more total pores and surface area, and thus a greater ability to hold water. Total water holding capacity will be greater. Thus, a silt loam can hold about 28% of its soil water in plant available form (between field capacity and the wilting point), whereas a sandy soil can hold perhaps 8% as plant available water[110].

Also highly important in the consideration of soil water is the ability of available, capillary water to move from storage areas in the soil mass to roots. As the roots deplete water films and pores in their vicinity, other areas nearby have a continuum of film surfaces along which

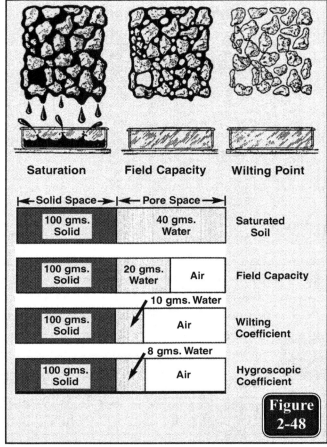

Figure
2-48

capillary water can flow. If soil structure is strong and contains numerous macropores with an interlacing continuum of water film surfaces, then capillary water flow along particle surfaces will be rapid, whereas a compacted and poorly structured soil, or a fairly imperious horizon, will retard the movement of water to root surfaces. See this contrast in Figure 2-50[111].

The facility of water movement through soils having different textures can be dramatically illustrated in the following two instances shown in Figure 2-51[112]. The sandy loam soil has many large pores which allow water to flow down quickly by gravity; capillary movement horizontally is restricted, since the water moves mostly down in a few hours. A

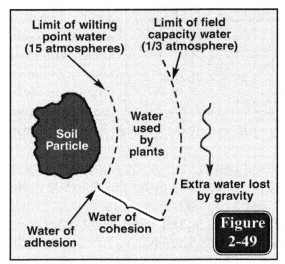

Figure
2-49

51

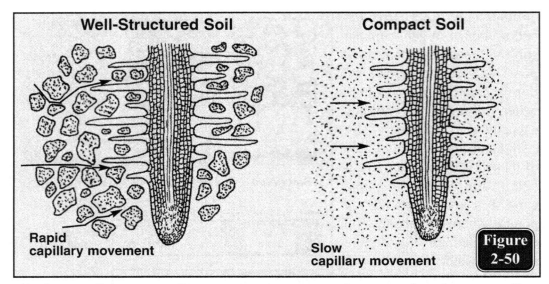

Well-Structured Soil

Compact Soil

Rapid capillary movement

Slow capillary movement

Figure 2-50

clay loam soil, by contrast, possesses many more micropores that slow water flow downward, allowing more horizontal flow by capillary action.

An impervious or highly porous layer in the root zone greatly alters water movement. A compact layer, of course, will slow water movement and allow anaerobic conditions to set in, killing desirable aerobic bacteria,

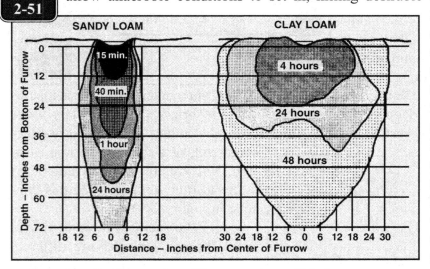

Figure 2-51

SANDY LOAM   CLAY LOAM

Depth – Inches from Bottom of Furrow

0
15 min.
12
40 min.
24
1 hour
36
48
24 hours
60
72

4 hours
24 hours
48 hours

18 12 6 0 6 12 18   30 24 18 12 6 0 6 12 18 24 30
Distance – Inches from Center of Furrow

fungi, and other organisms while allowing valuable nitrogen stores to denitrify and be lost. A sandy or gravely horizon beneath a clayey surface soil, on the other hand, will slow water movement downward until the heavier topsoil reaches field capacity. Then gravity will bring excess water into the coarser soil layers.

It is clear that water is a most critical component of the soil-plant system. It is the synergistic medium for the uptake of nutrients from the soil, for their transfer into the plant, and for the enzymatic reactions that occur within leaves, stems, and roots. Yet, water cannot function within living systems unless all other components are in place within the environment: soil mineral and organic matter, sunlight, air, and proper temperature. All of these components cooperate in a most profound and mutualistic manner. Each gives its contribution to life on a planet where the conditions are just right ... not that the absence of Satan's destructive forces would not make conditions much better. Even with this battle between life and death, the continual renewal of vitality of plants, animals, microbes, and all cellular life generation after generation speaks profoundly of the Architect of the earth's environment and the life forms that inhabit it. After all, the earth was meant to be inhabited (Isaiah 45:18), not to be a charred cinder in the outer reaches of space.

Light

While air, minerals, organic matter, and water are essential to life, the entire system is promoted and made possible by light energy from the sun. This is the only strictly outside import into the earth's biosphere that must be added to make plants grow. The energy from nuclear reactions is sent careening into space from the sun 93,000,000 miles away, reaching the earth in 8.3 minutes. The percentage of the sun's energy that actually intercepts the earth is extremely small, and some of the light that strikes it is reflected back out into space. Nonetheless, the light which plants eventually receive is sufficient for the growth we observe. Note Figure 2-52 for a summary of the many factors that influence light within the earth's atmosphere[113].

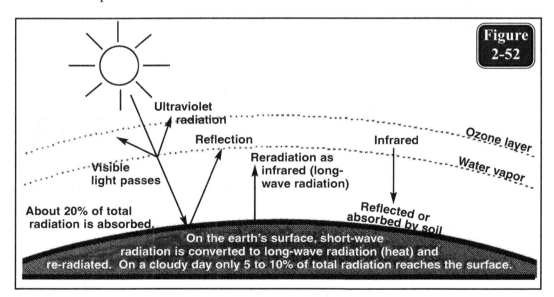

The amount of sunlight that strikes the earth at any particular location varies with the directness of the sun's rays and the degree of cloudiness. Much of the short-wave ultraviolet radiation is absorbed by the ozone layer, but the majority of visible rays reach the earth. Only a very small segment of the electromagnetic spectrum is visible, but within that spectrum many critical wavelengths power functions necessary for life (Figure 2-53). This is especially true for photosynthetic plants, but also is true for humans, who require sunlight on the skin to convert Vitamin D into its active form.

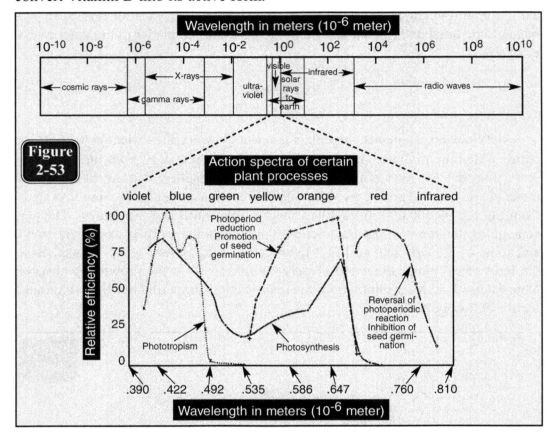

Notice in Figure 2-53 that photosynthesis (mainly chlorophyll a and b) uses light mainly in the violet-blue and orange-red wavelengths, allowing green to be reflected[114]. Phototropism requires violet and blue light; photoperiod reactions require yellow and orange for initiation and red for their reversal. Other photosynthetic pigments absorb light at different wavelengths from those utilized mainly for chlorophyll.

Light energy governs soil temperatures, plant photosynthesis and consequential carbon fixation and root oxidation, photosynthesis of autotrophic soil organisms such as cyanobacteria and algae, and rainfall to a large degree through its effects on evaporation, rainfall, and the entire hydrologic cycle. Truly, the sun's energy, the free gift of the Creator, powers all of life's systems on the earth, allowing all of the mutualistic elements, organisms, and biologic cycles to serve one another to produce the intricate interrelationships in the biosphere essential for life.

"Love your enemies, bless those that curse you, do good to them that hate you, and pray for them that despitefully use you and persecute you, that you may be children of your Father who in in heaven, for *He makes His sun to rise on the evil and on the good, and sends rain on the just and on the unjust[115]*".

Sunlight is a direct gift from the Father's sun. It represents a perfect mutualism and a pure example of love: life-giving power is sent to mankind and all living creatures without any thought of that power being returned. It is a totally selfless gift for which all of mankind ought to stand in awe ... not of the light itself, but of the One who made mankind and sustains him and the environment that supports them through this energy.

Soil Variations

The combined forces of parent material, vegetation, topography, climate, and time are unique for every locale on earth. Thus, the soils developed at every location are unique in some way, although they can be categorized into broad groups based on color, mineralogy, pH, depth, horizons, and certain other factors. A map of these categories across the United States is shown in Figure 2-54 along with a brief explanation of these different groupings[116].

Natural vegetation as influenced by rainfall and temperature dramatically affects soil development. With more rainfall, a greater leaf and root mass result that enriches the topsoil with organic matter to create mollisols, the very rich prairie soils of the eastern Plains and Corn Belt. Note the continuum of soils in Figure 2-55 from the short-grass prairies of northern Wyoming to the tall-grass prairies of southern Minnesota[117].

A= Alfisols. Soils with a gray to brown surface horizon, medium to high base supply, and subsurface horizons of clay accumulation; usually moist but may be dry during the warm season.

D= Aridosols. Soils with pedogenic* horizons, low in organic matter, and dry more than six months of the year in all horizons.

E= Entisols. Soils without pedogenic* horizons.

H= Histosols. Organic soils.

I= Inceptisols. Soils that are usually moist, with pedogenic* horizons from alterations of parent materials but not of accumulation.

M= Mollisols. Soils with nearly black, organic-rich surface horizons and high base supply.

S= Spodosols. Soils with accumulations of amorphous materials in subsur-

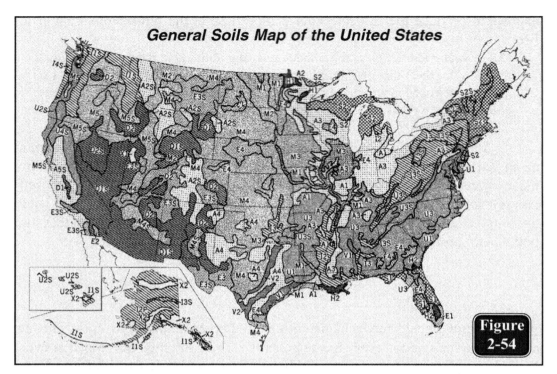

General Soils Map of the United States

Figure 2-54

face horizons.

U= Ultisols. Soils that are usually moist with horizons of clay accumulation and a low base supply.

V= Vertisols. Soils with a high content of swelling clays and wide, deep cracks at some season.

X= areas with little soil.

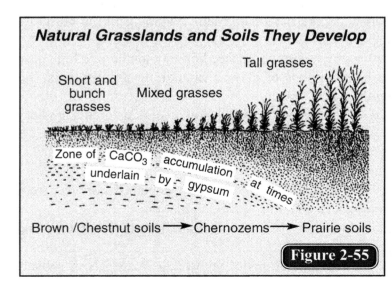

Natural Grasslands and Soils They Develop

Tall grasses

Short and bunch grasses

Mixed grasses

Zone of CaCO₃ accumulation at times underlain by gypsum

Brown /Chestnut soils ➔ Chernozems ➔ Prairie soils

Figure 2-55

Pedogenic =processes that lead to the development of defined horizons or layers in the soil that differ in color, chemical traits, or physical conditions.

The five factors of soil formation cooperate to form definite soil horizons in most locations. These horizons may be generalized as in Figure 2-56, called a *soil profile*[118].

Some pictures of soil profiles from different locations are shown in Figures 2-57, 2-58, and 2-59[119]. They have all developed

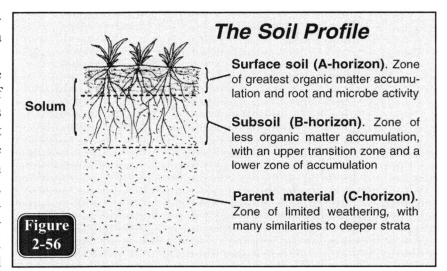

The Soil Profile

Solum

Surface soil (A-horizon). Zone of greatest organic matter accumulation and root and microbe activity

Subsoil (B-horizon). Zone of less organic matter accumulation, with an upper transition zone and a lower zone of accumulation

Parent material (C-horizon). Zone of limited weathering, with many similarities to deeper strata

Figure 2-56

from forces of parent minerals, organisms and organic matter, rainfall and temperature, and slope position over time.

| Figure 2-57 | Figure 2-58 | Figure 2-59 |

Forest Soil

Eastern Texas

Developed on marine very fine sand

A-horizon shallow, highly acidic, low in organic matter

Poorly mineralized

Prairie Soil

East-Central Illinois

Developed on glacial till and alluvium

A-horizon slightly acidic, deep, high in organic matter

Highly mineralized

Steppe Soil

Central Wyoming

Developed on limestone bedrock

A-horizon alkaline and thin but with some organic matter

Highly mineralized

Chapter III
How the Soil Works With Plant Roots

Having briefly covered the nature of plants and soils, we now can deal with the functioning of the plant and organisms in the soil. None of them can be discussed separately, since each serves the others within a highly dynamic system It is imperative to understand that, besides providing anchorage in the soil, roots are primarily essential for taking up nutrients and water for above-ground growth. The following discussion will cover in some detail the essential chemical processes involved with nutrient uptake. A few of these processes involve the uptake of complex macromolecules, such as by phagocytosis. The biological processes which convert unavailable nutrient forms to mobile and usable forms will be covered after this section, for both rhizosphere and non-rhizosphere soil. While the real workhorses in the nutrient uptake scheme are the soil microorganisms, they rely integrally on chemical forces in nature to expedite their nutrient processing and delivery to plant roots.

Plant Nutrient Uptake Mechanisms

The uptake of plant nutrients from a soil environment is a vastly complex phenomenon involving not simply the plant root, but a wide array of microorganism species which directly feed upon root exudates. During the life of a plant up to 25% or more of the plant's chemical energy, in the form of carbon compounds which are manufactured in the leaves, is lost into the soil directly adjacent to the roots[1]. The microorganisms — fungi (especially mycorrhizae), bacteria, algae, actinomycetes, and others — help supply nutrients to the growing root through processes which will be described.

The nutrient uptake process by plant roots in natural soil environments may be divided into three phases: (1) the movement of nutrients from the soil to the root (along with root growth into the soil), (2) absorption of these nutrients across cell membranes, especially across the Casparian strip barrier[2], and (3) uptake by vascular elements. These three phases are illustrated in Figure 3-1[3].

Only the first two phases will be dealt with at length. A diagrammatic view of this process is shown in Figure 3-2, where ions, indicated by dots, are concen-

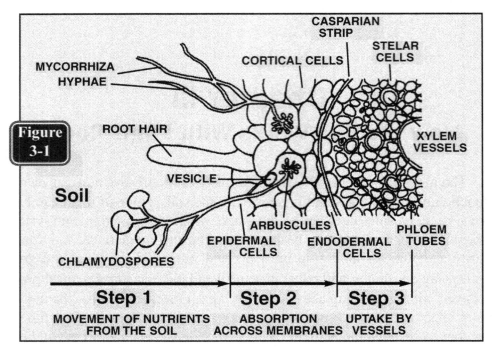

Figure 3-1

trated outside the root; they then enter the "outer space"[4]. This space is comprised mostly of the cell walls of the epidermal and cortical cells, inward to the endodermal cells (Casparian strip), which constitutes a barrier to ion diffusion. Epstein indicated that transport mechanisms may bring ions into the cytoplasm ("inner space") of epidermal and cortical cells as well[5]. At this point the nutrient ions not already within the cytoplasm must enter the cytoplasm ("inner space") of the endo-

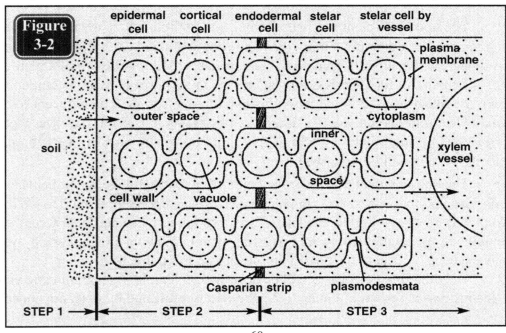

Figure 3-2

dermal cells through transport mechanisms, and travel finally from cell to cell through the plasmodesmata to the vascular tissues[6]. Nutrient enrichment of the solution moving through the tissues is usually very noticeable on the inner side of the endodermal cells[7].

Movement of Nutrients from the Soil to the Root

Nutrient supply to the root surface may be divided roughly into three types of processes[8]:

a. **Root interception** or extension: roots grow into the soil to intercept exchangeable nutrients, and take them up through **contact exchange.**

b. **Mass flow** of ions with the soil solution resulting from transpirational flow

c. **Diffusion** of ions toward the root surface when uptake is faster then supply

It should be realized that these three processes, which act in concert, do not make up the sum total of mechanisms of nutrient uptake, since a major source of uptake, that mediated by mycorrhizae, may in some cases overshadow all of the others. Since nearly all important agronomic crops contain mycorrhizae (usually the vesicular-arbuscular type[9]), and these fungi may increase the effective absorbing surface of a host root by as much as 10 times[10], the value of such microorganisms to the host plant must be recognized.

It should also be realized that several factors affect rates of root nutrient uptake, such as soil pH, salt concentration of complementary ions, microorganisms present, plant species, and soil bulk density[11]. Additional factors include soil conditions such as temperature, light, soil moisture content, aeration, the presence of toxic elements or herbicides, and nutrient sufficiencies or deficiencies[12].

Different ions are taken up in different proportions by these mechanisms as shown in Table 3-1[13]. Sodium uptake, for instance, is highly related to root interception and mass flow, while most of the K is supplied by diffusion[14]. Calcium is taken up almost entirely by root interception, and also Mg, though for Mg mass flow and diffusion are also important on certain soils. Others have confirmed this data by stating that elements such as K, NO_3-N, or SO_4-S are relatively mobile and move with the soil solution to the root largely by diffusive processes[15]. Calcium and Mg, however, due to tighter binding to the soil colloid, are absorbed more through root interception. Innate ion mobility in the soil is thus shown to be extremely important in regard to the mode of root uptake. Because the mycorrhizae can proliferate within such a vast volume of soil, with many times the absorbing surface than the roots, their value in absorbing immobile ions firmly attached to clay and organic matter colloidal particles can be considerable[16].

Actual movement of ions to roots within the soil occurs over very small distances; mycorrhizae can move nutrients much further[17]. Note Figure 3-3, showing also the existence of a mucilaginous layer alongside the root cell wall[18]. This

Correlation of uptake from various supply modes with total uptake by *Avena fatua*[11]		Na	K	Ca	Mg
Root interception	Surface	0.920***	0.507**	0.962***	0.425**
	Subsoil	0.565***	0.538***	0.996***	0.398*
Mass flow	Surface	0.929***	0.552***	-0.043	0.596***
	Subsoil	0.985***	0.387*	-0.241	0.617***
Diffusion	Surface	0.012	0.800***	I	0.778***
	Subsoil	-0.0222	0.583***	I	0.591***
Accumulation	Surface	0.925***	I	0.641***	-0.181
	Subsoil	0.924***	I	0.789***	-0.152
Interception +	Surface	0.990***	0.560***	1.000***	0.649***
Mass flow	Subsoil	0.999***	0.412*	1.000***	0.824***

*Significant at P= .05; ** significant at P= 0.1; ***significant at P= .001; I = insufficient non-zero data.

Table 3-1

layer has been commonly observed on root surfaces [19] and serves as a focal point for vigorous microbial activity and nutrient release. Contact exchange, where the root actually contacts soil colloidal particles and excretes hydrogen ions for

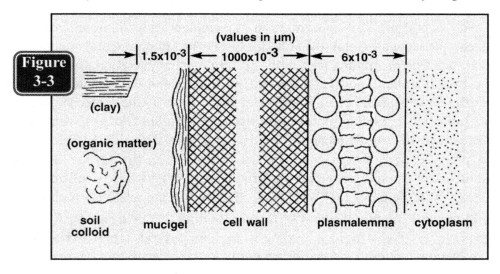

Figure 3-3

(values in µm)

1.5×10^{-3} 1000×10^{-3} 6×10^{-3}

(clay)

(organic matter)

soil colloid   mucigel   cell wall   plasmalemma   cytoplasm

exchange with nutrient cations such as Ca, Mg, Cu, Mn, or others, can be an important process during root extension ... especially for the uptake of immobile nutrients [20]... although with less than 1 percent of the soil's volume contacted by root hairs [21] nutrient uptake by contact exchange may at times be overestimated.

Mycorrhizal hyphae may greatly increase the effective absorbing surface of the root and increase nutrient uptake [22]. Their hyphae can extend up to 8 cm into the surrounding soil and transport nutrients that full 8 cm back to the roots[23], tremendously extending the limited zone of nutrient extraction along root hairs.

The soil colloids (organic matter and clay) form a dynamic and extensive continuum in the soil, so their contact relations are of great significance[24]. As soil microorganisms break down organic matter and release nutrient ions, or as fertilizer elements upset the equilibrium of the clay-organic matter complex, cation exchange reactions release nutrient elements into the soil solution which may then be intercepted by the extending root or absorbed by the root through mass or diffusive flow. This activity becomes especially intense within the rhizosphere where root exudation, and thus highly active microorganism growth, is occurring.

During root extension, root surface "fibrils" appear to invade organic colloids, a phenomenon likely also to occur in clays[25]. Particles or organic debris at the root tip are integrated with rhizoplane fibrils at regions where ion uptake is localized, and where root surface area can be greatly extended by the fibrils. When the diffuse double ion layer is exposed to a sudden change in cation concentration, as when the fibril approaches, the ions "bleed off" the double layer and cations are redistributed to reestablish an equilibrium; see these events in Figure 3-4[26].

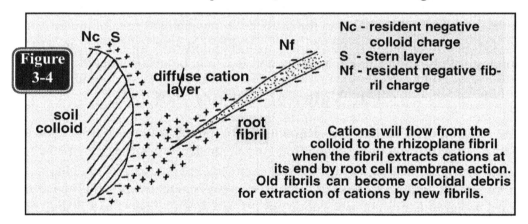

Figure 3-4

Nc S — Nf

Nc - resident negative colloid charge
S - Stern layer
Nf - resident negative fibril charge

diffuse cation layer

soil colloid

root fibril

Cations will flow from the colloid to the rhizoplane fibril when the fibril extracts cations at its end by root cell membrane action. Old fibrils can become colloidal debris for extraction of cations by new fibrils.

Mass flow and diffusion of nutrients from the soil into the soil solution, and then into the root, may be summarized by the following simple scheme[27] (Figure 3-5). M indicates any nutrient, and ML a complexed form which may be translocated to the root surface.

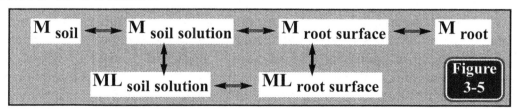

M soil ⟷ M soil solution ⟷ M root surface ⟷ M root

ML soil solution ⟷ ML root surface

Figure 3-5

The solute transport mechanisms of mass flow and diffusion are shown in

Figure 3-6[28].

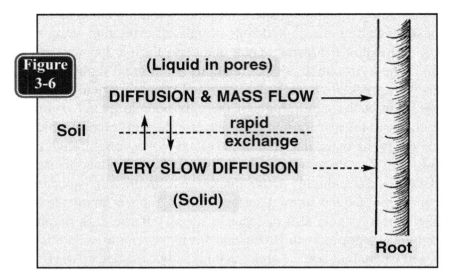

The points above illustrate the rapid, dynamic equilibrium between solutes in the soil pore solution and ions adsorbed on adjacent soil surfaces. The adsorbed ions tend to buffer the soil solution against changes in concentration induced by root absorption[29]. According to the mass flow mechanism several nutrients (Ca, Mg, Na, and Mo) can be supplied to a crop like corn for maximum growth, but supplies of P, NO_3-N, Cu, Fe, Mn, Zn, and B would be insufficient[30].

The Value of Mycorrhizae

Based on what has been learned above, the importance of additional supplies of nutrients provided through microorganism activity in the rhizosphere, plus the transport of nutrients by mycorrhizae from beyond the reach of root hairs, cannot be overemphasized. Vesicular arbuscular (VA) mycorrhizal fungi have been shown to stimulate plant absorption of P[31], Zn[32], Ca[33], Fe[34], Mg[35], and Mn[36]. These mutualistic fungi are discussed later.

The VA fungi are short-lived, surviving less than two weeks before being digested by the host plant[37]. The host plant usually releases large amounts of exudates into the root zone[38], so by releasing more or less exudates the plant can regulate the abundance of mycorrhizal colonization[39].

Higher fertilizer requirements to produce the same crop yield in recent years compared to needs in past years could relate to the destruction or suppression of mycorrhizal populations by biocides or pesticides[40], topsoil removal, or planting non-mycorrhizal crops[41]. In citrus orchards mycorrhizal fungi may be equivalent to 100 to 500 pounds per acre of P when contrasted to the massive P applications required for fumigated soil or hydroponic solutions[42].

A Key to Plant Nutrient Uptake: Rhizosphere Activity and the Oxygen-Ethylene Cycle

At the heart of activities of the soil-root interface lies the rhizosphere, the zone of intense biological activity which extends only a few mm out from the root. A plant may excrete up to 25 percent or more of its chemical energy, manufactured in the leaves, out through the root into the rhizosphere ... as exudates or as dead plant cells[43]. In solution cultures usually less than 2 percent of root carbon is released as exudates[44]. The plant goes to this trouble for good reasons, since such an energy-costly activity bears dividends by feeding microorganisms that perform complex nutrient-supplying functions[45]. The mycorrhizal fungi, already discussed, belong to this group of organisms benefitting from these exudates, as do *Rhizobium* bacteria, cyanobacteria, *Azotobacter*, and certain others. Grass-bacteria associations in tropical soils benefit from the root-supplied substrate for nitrogen fixing activity of the bacteria; for bahia grass [*Paspalum notatum* Flugge] up to 5.0 kg/ha can be fixed over three months, though great differences occur between cultivars[46].

The compounds in the exudate of plant roots serve as food for microorganisms in the rhizosphere. When plants are stimulated by certain substances such as growth regulators, antibiotics, and certain foliar or soil-applied fertilizers, the root hairs may be stimulated to excrete more of these carbon compounds, or alter the composition of exudates[47]. The growth of microbes in the rhizosphere leads to oxygen depletion, the creation of anaerobic microsites, and ethylene ... these processes forming the nucleus for anionic and cationic nutrient release from soils through the oxygen-ethylene cycle[48], as will be discussed later.

Absorption of Nutrients Across Membranes

Once nutrients have reached the root surface by diffusion or mass flow of the soil solution, by contact exchange, through release or synthesis of organic compounds by microorganisms, or by transport from more remote locations by mycorrhizal fungi, they must somehow enter the root. These processes are not totally understood, though in most cases it is known that energy must by utilized to "pump" in the nutrients against a gradient of higher ionic concentration within the cytoplasm of the plant root cell[49].

The possible mechanisms of ion uptake by plant roots — and all types of cells — are shown in Table 3-2[50]. Some of the same principles are involved in root uptake as for the movement of ions from the soil to the root surface. Not every mechanism will be covered directly in this section, though all of them are interrelated and thus will be at least indirectly addressed.

Possible Ion Uptake Mechanisms in Biological Systems

Mechanism	Ion accumulation	Ion selectivity	Principle	Probable distribution
Simple diffusion	-	-	molecular movement	universal
"Facilitated diffusion"	-	+	polypeptide carrier	microorganisms
"Mass flow"	+	-	transpiration	universal in vascular land plants
Donnan equilibrium	+	+	biopolymer synthesis	universal
Redox pump	+	+	electrophoretic transport	yeast, animals
Chemi-osmosis	+	+	charge separation	mitochondria, chloroplasts
Anion respiration	+	+	cytochrome-dependent	mitochondria?
"Carrier complex"	+	+	ATP-dependent protein movement	tonoplast?

Table 3-2

There is some question as to whether inorganic ions alone — not higher molecular weight organic molecules — are able to traverse cell membranes and enter the transpiration stream. Many plant scientists believe that only inorganic ions are taken up[51], or at the very most small organic ions. Yet, there should be little doubt any longer that roots are capable of absorbing organic macromolecules with the advent of systemic soil-applied herbicides and other pesticides[52]. Picloram uptake by oat and soybean roots is likely due to both passive and active mechanisms[53], while 2, 4-D uptake most likely involves an active root mechanism[54]. Many other studies with other pesticides have shown similar results. There is also the likelihood that phagocytosis is a significant means of nutrient uptake of macromolecules by plants.[55] In Japan considerable enthusiasm has been generated by some workers for mechanisms whereby plants directly utilize molecules of high molecular weight[56].

Since most studies in nutrient uptake by root cells have involved simple ions, often in solution culture, most of the work presented in this section will reflect this bias. Most authors indicate that the forms of nutrients taken up are NO_3-N and NH_4^+ (nitrogen), $H_2PO_4^-$ and HPO_4^{-2} (phosphorus), K^+ (potassium), Ca^{+2} (calcium, Mg^{+2} (magnesium), SO_4^{-2} (sulfur), BO_3^{-3}, $B_4O_7^{-2}$, or $H_2BO_3^{-3}$ (boron), Fe^{+2} or as a complex organic salt (iron), Mn^{+2} or complexed -Mn (manganese), Cu^{+2} or complexed Cu (copper), Zn^{+2} or complexed Zn (zinc), MoO_4^{-2} (molybdenum), Cl-(chlorine), and Na^+(sodium)[57]. No mention is made in this or other basic texts of organic macromolecule uptake[58].

Cell Membrane Structure

To better understand nutrient uptake into cells it is necessary to understand the nature of cell membranes through which nutrients must pass. Any proposed membrane model must account for the following[59]:

1. A high permeability to water, the rate of penetration being much greater than for solute molecules.
2. Permeability to non-electrolytes is generally correlated with lipid solubility of the solute.
3. Electrical resistance is high, normally 10^9 to 10^{11} cm.
4. The surface tension values (0.1 to 2.0 dynes/cm), probably indicating a protein layer at the exterior.
6. They are frequently lysed by treatment with lipid solvents or proteolytic enzymes.

Among the earliest proposed membrane models was that of Danielli and Davson[60], who devised a double layer protein-lipid model which has been only slightly modified over the years. Others have attempted to improve on the model[61]. Models of Danielli and Davson and Lenard and Singer[62] are shown in Figure 3-7.

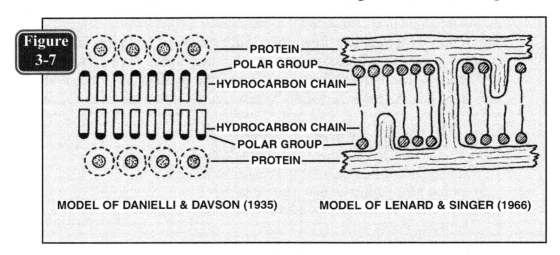

Figure 3-7

PROTEIN
POLAR GROUP
HYDROCARBON CHAIN
HYDROCARBON CHAIN
POLAR GROUP
PROTEIN

MODEL OF DANIELLI & DAVSON (1935)   **MODEL OF LENARD & SINGER (1966)**

While most membranes consist almost entirely of protein and lipid, there is still considerable variation in the proportion of each for different types of cells and organelles. Yet, a protein fraction containing phospholipids, sterols, glycolipids, and some triglycerides indicates a somewhat common organization among different cells and organelles[63]. It is within the phospholipid complex that nutrient uptake mechanisms are found.

It is possible that "tunnel proteins" may help explain ion transport through membranes[64]. The ion, or ion-carrier complex, might simply slide along the protein molecule to the other side of the membrane. *Ionophores*, molecules having a hydrophobic exterior and a hydrophilic interior, change in conformation depending on the composition of the medium and may explain how ions can pass through a membrane[65]. Thus, an ion like K^+ could dissolve in the water-soluble interior of the molecule, be transferred with the molecule as it passes through the nonpolar hydrocarbon layer of the membrane, be released, and then the molecule could return to repeat the process.

Active and Passive Uptake

The similarity of active (mediated) ion uptake by root cells to enzyme activity is quite profound. For this reason the principles of enzyme kinetics have been applied to nutrient uptake through cell membranes ... thereby imputing enzyme systems in the actual uptake scheme. There are two basic means by which solutes may cross membranes[66]:

1. **Nonmediated (passive) transport**. The rate of transport is always directly dependent on the solute concentration, and its temperature coefficient is usually that of physical diffusion (about 1.4 per 10° C rise in temperature). This form of transport is entirely the result of simple physical diffusion of the solute alone in response to a concentration or electrochemical gradient. The solute molecule is neither chemically modified nor associated with another species in its passage through the membrane. The system loses free energy. Ion uptake is proportional to the gradient of concentration across the plasmalemma, basically following Fick's first law of diffusion[67]:

$$D = K\frac{s}{t} \text{ , where } K = \frac{x}{a(C_1-C_2)}$$

x = membrane thickness

a = area

C_1-C_2 = concentration gradient

D = specific diffusivity of the substance in the cell, or the cell's permeability to the substance
s = amount of the substance diffusing, in moles
t = time, in seconds

2. **Mediated (facilitated or active) transport**. This type of membrane transport may be either passive or active, but above all it displays saturation kinetics: the transport system can become saturated with the substance transported, just as enzymes can become saturated with their substrates. The system gains free energy.

The two systems of nutrient uptake by cell membranes are diagrammatically described in the Figure 3-8.

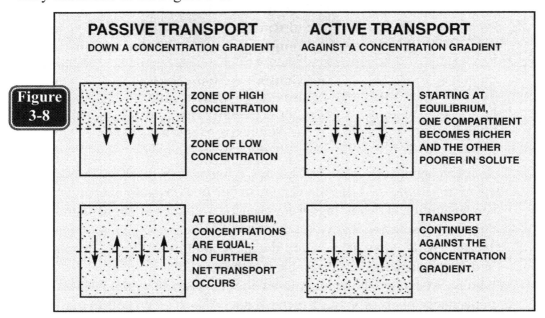

Figure 3-8

PASSIVE TRANSPORT
DOWN A CONCENTRATION GRADIENT

ZONE OF HIGH CONCENTRATION

ZONE OF LOW CONCENTRATION

AT EQUILIBRIUM, CONCENTRATIONS ARE EQUAL; NO FURTHER NET TRANSPORT OCCURS

ACTIVE TRANSPORT
AGAINST A CONCENTRATION GRADIENT

STARTING AT EQUILIBRIUM, ONE COMPARTMENT BECOMES RICHER AND THE OTHER POORER IN SOLUTE

TRANSPORT CONTINUES AGAINST THE CONCENTRATION GRADIENT.

Mediated transport kinetics resemble enzyme kinetics because saturation kinetics are noted, substrate specificity is shown, and inhibitors can frequently specifically inhibit transport[68]. Together, these three points strongly suggest that plant membranes contain molecules capable of reversibly binding specific substrates and transporting them across membranes. These proteins, or protein systems, have been called **transport systems**, **carriers**, **porters**, **or translocases**[69]. The carrier molecules may rotate within the membrane, carry their ligands across the membrane by diffusion, or undergo conformational changes to create a "hole" in the membrane for the substance to be transported. Mediated transport is sometimes called **facilitated diffusion** in recognition of the fact that all transport systems, mediated or not, ultimately result from diffusion; even in mediated transport a substrate-carrier complex responds to a gradient of the complex within the membrane[70].

Classical Michaelis-Menton transformations in enzyme-substrate reactions [A + B ⟷ AB] are shown in the following chemical equations[71].

Without going into details of the transformations (most modern biochemistry texts discuss them), the equation below is obtained, from which a Michaelis-Menton plot may be drawn.

Further evidence that active ion transport across membranes occurs is given in the following six points[72]:

1. Accumulation ratios: Salts inside cells may reach far higher concentrations

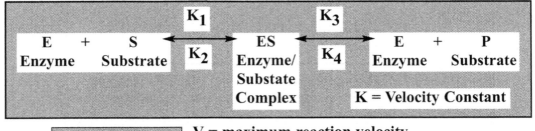

$$V = \frac{v \max [S]}{K_m + [S]}$$

V = maximum reaction velocity
[S] = substrate concentration
v = rate of enzymatic reaction
K_m = the Michaelis-Menton constant

than levels outside cells. For K^+ or Cl^- this may be 1,000: 1 or higher.

2. *Temperature effects:* The Q_{10} (change in rate of a physiological process) for ion absorption is two or higher for each 10°C change in temperature. Near freezing, absorption is severely inhibited.

3. *Oxygen:* At low oxygen tensions the absorption of many ions is inhibited, showing a dependence on aerobic metabolism.

4. *Poisons:* Many metabolic inhibitors will inhibit ion absorption, showing the dependence of absorption on metabolism. Also, a given poison may inhibit the transport of two ions of a salt quite differently.

5. *Carbohydrates:* Sugar reserves are often positively correlated with ion accumulation (This may relate to root exudation and stimulation of rhizosphere microflora).

6. *Light:* Light produces metabolites necessary for ion absorption, and thus promotes it.

The actual mechanism by which active transport occurs is becoming slightly clearer[73]. Carriers have been postulated for a long time, and are now visualized as compounds that complex the ion and render it more "soluble" to the membrane, such as the specific alkali-metal binding cyclic compounds[74].

Models have been proposed for explaining the active transport of ions across membranes[75,76,77]. Note Figures 3-9 and 3-10, which show how ions may be moved through the cell membrane into the cell using a carrier of some sort. Proteins rank high on the list of likely carriers, amongst other compounds.

It has been discovered that there are actually two active transport mechanisms, one operating at low ion concentrations (Mechanism I) and the other at higher concentrations (Mechanism II)[78]. Mechanism I follows simple Michaelis-Menton kinetics while Mechanism II shows a more complex pattern, suggesting that the carrier involved has several active sites which differ in their affinity for the ion[79]. In cells with vacuoles both mechanisms have been found to operate, and in cells having no vacuoles only Mechanism I operates (in the plasmalemma)[80].

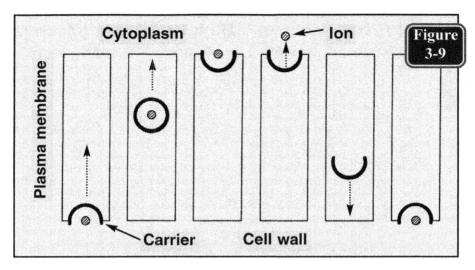

Figure 3-9

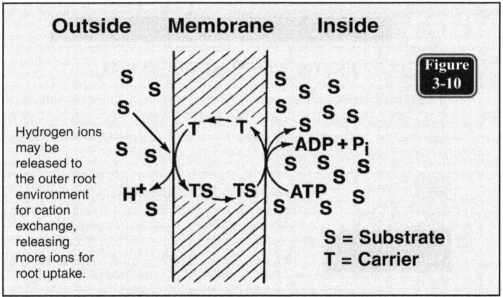

Figure 3-10

Hydrogen ions may be released to the outer root environment for cation exchange, releasing more ions for root uptake.

S = Substrate
T = Carrier

Apparently these mechanisms help a plant operate under different environmental conditions: thus, a salt tolerant plant has a low Km for Mechanisms I and II, while a salt intolerant plant has a low Km for Mechanism I but a high Km for Mechanism II and will die of plasmolysis in a high salt environment. Note Figure 3-11 for a model of so-called "allosteric control" of ion flow through a cell membrane — in this case for potassium flowing into barley root cells — possibly explaining the operation of Mechanisms I and II.

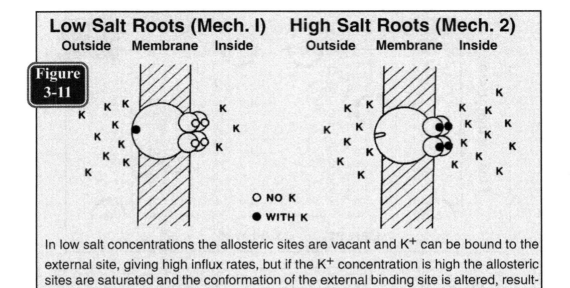

Low Salt Roots (Mech. 1)
Outside   Membrane   Inside

High Salt Roots (Mech. 2)
Outside   Membrane   Inside

Figure 3-11

○ NO K
● WITH K

In low salt concentrations the allosteric sites are vacant and K^+ can be bound to the external site, giving high influx rates, but if the K^+ concentration is high the allosteric sites are saturated and the conformation of the external binding site is altered, resulting in a lower affinity for K^+.

Phagocytosis of Macromolecules

As mentioned earlier, phagocytosis of macromolecules is considered to be a major means of nutrient absorption by several scientists, especially some from Japan. A number of means may be used by a root cell to absorb nutrients by invagination of the cell membrane (see the Figure 3-12)[81].

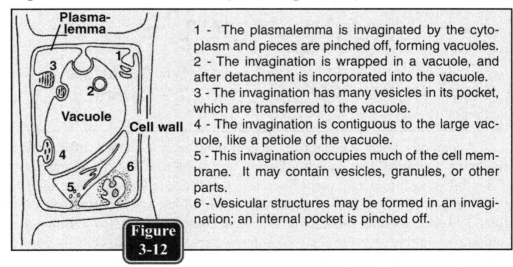

Plasma-lemma

Vacuole

Cell wall

1 - The plasmalemma is invaginated by the cytoplasm and pieces are pinched off, forming vacuoles.
2 - The invagination is wrapped in a vacuole, and after detachment is incorporated into the vacuole.
3 - The invagination has many vesicles in its pocket, which are transferred to the vacuole.
4 - The invagination is contiguous to the large vacuole, like a petiole of the vacuole.
5 - This invagination occupies much of the cell membrane. It may contain vesicles, granules, or other parts.
6 - Vesicular structures may be formed in an invagination; an internal pocket is pinched off.

Figure 3-12

One source said, "... since J. von Liebig established the 'mineral theory of nutrition' in 1840, the capacity of roots to absorb organic molecules as important plant nutrients has been ignored by many investigators"[82]. As early as the 1950's it has been shown that macromolecules can enter root tip cells through the use of ribonuclease[83]. Later it was reported that enzymes and other proteins were absorbed by barley roots and suggested pinocytosis as the uptake mechanism[84]. Other workers (Coulomb, 1973; Mahlberg et al., 1974; Robards and Robb, 1974; Sung and McLaren, 1975) have reported the detection of plasmalemma invaginations and pinocytosis, some using radioactive labeled molecules such as albumin[85]. At least a portion of the macromolecules were taken up without degradation. According to Robards and Robb, "Pinocytosis as a factor in uptake of materials by plants should not be completely ignored in the future as it has been in the past by most investigators"[86].

Transport into Vascular Tissues from Root Tissues

It is not the object of this discussion to delineate at length the delivery system of solutes from the cells of the root to the vascular system. However, there has been knowledge gained in this area recently to partially reveal how uptake occurs. Within the cytoplasm of contiguous cells the solutes move in a very orderly fashion, primarily through tiny pores which connect cells called plasmodesmata[87]. The ions are absorbed within the "inner space" (cytoplasm) at or near the root surface, and then move by symplastic flow across the cortex, endodermis, and pericycle within cells, being passed by the plasmodesmata from cell to cell (Figure 3-13)[88].

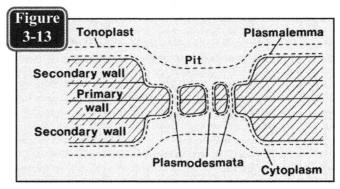

Figure 3-13

Tonoplast — Plasmalemma — Pit — Secondary wall — Primary wall — Secondary wall — Plasmodesmata — Cytoplasm

Despite considerable work and review of research in this area during past years "there is as yet no agreement on the fundamental processes whereby ions are transferred into the xylem of plant roots, usually against a concentration gradient, and there is still uncertainty on the exact sites of the various processes and on the path followed by ions"[89].

Non-Rhizospheric and Rhizospheric Soil

There are two basic environments within the soil: the soil mass outside the root zone (non-rhizospheric soil), and the soil within the close vicinity of root surfaces (rhizospheric soil). In either case, the soil organisms respond to the food supply available. If plant residues are added to the soil surface the fungi and bacteria multiply to consume it and break it down to humic by-products to benefit the soil physical and chemical status. If exudates are emitted along the root surface, organisms multiply to consume them and produce by-products to benefit the plant. These symbiotic interrelationships are powerful and pervasive.

Non-Rhizospheric Soil

The array of organisms found in soils has already been discussed. These include a wide variety of fungi, bacteria, algae, cyanobacteria, actinomycetes, protozoa, nematodes, microarthropods, earthworms, millipedes, ants, and various other insects and arachnids. Each occupies a particular niche within the soil depending upon nutrients available, aeration, temperature, and other variables. They are found in both rhizospheric and non-rhizospheric soil, and will multiply wherever conditions are proper for them ... lying in wait with armies of billions and trillions of spores and active cells.

Organisms will proliferate when organic or inorganic substances are available to serve as food. Thus, decaying roots within the soil matrix, fallen stems on the soil surface, or manure from livestock will provide a haven for greatly accelerated growth and proliferation of fungi and bacteria. All of these foods for the microbes originate from plants. The organisms that attack and break down the organic matter follow in a succession. Usually fungi first attack the fresh material. Then other species of fungi will feed on the residues left by the first feeders. One serves the other, even if a generation must die to feed the next. Bacteria and actinomycetes move in, as well as protozoa, earthworms, mites, and nematodes. All of these organisms require a fixed carbon source for their existence: they are thus called **heterotrophs**. Once the energy is used up the microbes will either form spores — a thick-walled, water and temperature resistant stage which allows for long periods of hibernation until favorable conditions for growth return — or live at a much less vigorous level. Humic substances, the semi-resistant end product of the initial breakdown of raw organic materials, provide enough energy for a continuously operating microbial population at low levels. The intensity of that level depends on how frequently organic residues are returned to the soil. A possible scheme for successional organism breakdown is shown in Figure 3-14[90].

The dark, incoherent, heterogeneous colloidial material from the microbial breakdown of vegetation and manures is called **humus**. The enzymes of fungi, bacteria, and other organisms break the original plant tissue into this dark residue that is fairly resistant to further decomposition; see Figure 3-15 for an electron micrograph of decomposing leaf litter[91]. Compounds are either modified or synthesized by the microbes. If breakdown of the plant tissue would be carried to completion, all that would remain would be minerals, the same constituents that first were taken up by the plants from the soil in the process of tissue synthesis.

Certain of the breakdown products of plant tissue have specific growth-

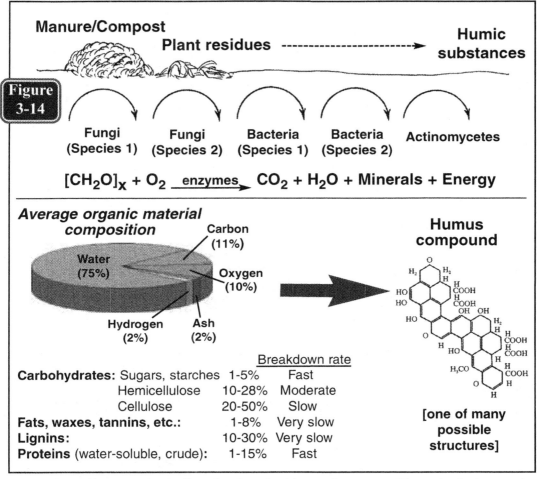

Figure 3-14

Manure/Compost
Plant residues - - - - - - - - - - - - - - - - - - - → **Humic substances**

Fungi (Species 1)   Fungi (Species 2)   Bacteria (Species 1)   Bacteria (Species 2)   Actinomycetes

$$[CH_2O]_x + O_2 \xrightarrow{\text{enzymes}} CO_2 + H_2O + \text{Minerals} + \text{Energy}$$

Average organic material composition

Water (75%)
Carbon (11%)
Oxygen (10%)
Hydrogen (2%)
Ash (2%)

Humus compound

[one of many possible structures]

	Breakdown rate	
Carbohydrates: Sugars, starches	1-5%	Fast
Hemicellulose	10-28%	Moderate
Cellulose	20-50%	Slow
Fats, waxes, tannins, etc.:	1-8%	Very slow
Lignins:	10-30%	Very slow
Proteins (water-soluble, crude):	1-15%	Fast

promoting effects, and are directly absorbed intact by roots. Those include certain vitamins, growth regulators (such as cytokinins and auxins), hormones, and organic acids such as humic acids that can be absorbed and moved into the plant to stimulate growth processes[92]. This issue will be touched upon again later.

Notice how the humus in Figure 3-14 contains a large number of "functional groups", such as COOH and OH. These sites ionize in soil water to create exchange sites for cationic minerals, showing why soil organic matter contributes so greatly to cation exchange capacity ... even more than do clay minerals.

The total mutualistic cycle of carbon moving through the soil illustrates how powerfully the growing crop — nourished by the sun — feeds the soil

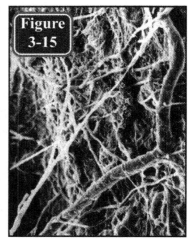

Figure 3-15

organisms through root exudates, which organisms degrade the fresh, unstable organic compounds to stable humic substances. As minerals are gradually released from the humus through enzymatic breakdown, roots take up these nutrients and flourish at the selfless contribution of the previous generation's death.

The carbon cycle is reviewed in Figure 3-16[93].

Besides organisms that require organic matter for growth in the soil, some organisms contain chlorophyll and can photosynthesize and thus fix their own carbon from CO_2 in the air. These are called **autotrophs.** Some can even fix nitrogen as well, and a few — notably the cyanobacteria (blue-green algae) — can fix both carbon and nitrogen (see Figure 3-17)[94].

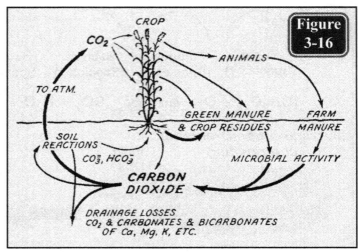

Figure 3-16

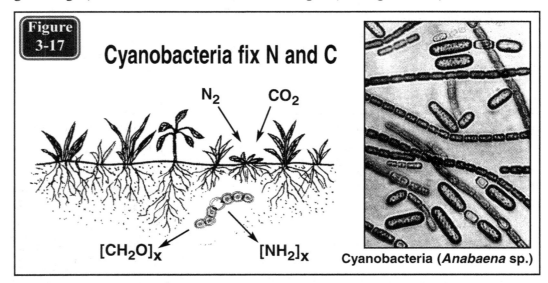

Figure 3-17

Cyanobacteria fix N and C

Cyanobacteria (*Anabaena* sp.)

Normally these autotrophic organisms live at the soil surface where light is available. Cyanobacteria, however, have been shown to colonize the rhizosphere as well where root exudates are available. They have even been found in the stomata of leaves and within the leaf itself, apparently fixing nitrogen for plant use directly where it is needed on what is called the **phylloplane.** The blue-green sheen noted on moist soil surfaces in the summer is most likely a covering of cyanobacteria. It is even possible that solar energy can be piped down the plant's stem into roots like fiber optics, granting photosynthetic energy to cyanobacteria and other photosynthesizing organisms within the rhizosphere.

The Nitrogen Binding Character of Soil Organisms

While the cation exchange sites of clay and organic matter bind cations (Ca^{++}, Mg^{++}, K^+, Na^+, etc.) in the soil, anions are thought to be held by a complementary **anion exchange capacity** that may originate within the organic matter and clays. **Anions**, or negatively charged elemental forms, must attach to positively charged sites, or else in some way be bound through the growth of crystals on mineral surfaces or through other forces. The fact remains that mechanisms operate within the soil to hold anions like SO_4^{-4}, $H_2PO_4^-$, NO_3^-, MnO_4^-, BO_3^{-3}, and others against leaching.

Soil bacteria and fungi perform this binding function better than any other

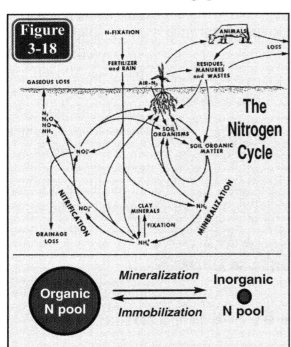

mechanism, especially for nitrogen. Nitrogen can easily be lost through leaching, denitrification, or erosion. Though the nitrogen cycle has already been discussed, it is important to examine the cycle to appreciate avenues for potential loss of this critical nutrient (Figure 3-18)[95].

At any one time most of the soil nitrogen is stored within the organic matter (Figure 3-18, bottom). It is released at a slow rate by microbial enzymatic processes, ideally at the rate plants require nitrogen ... since optimum temperature and moisture conditions for plant growth are also the optimum conditions for soil microbial activity. Not only is CO_2 released as enzymes break down the organic materials, but associated nitrogen and other minerals that constitute the organic fabric are released as well.

Notice how the soil, air, and sunlight form an intensive community that strives to provide plants with the necessary nitrogen for vigorous growth. Each contributes its part, interacting with the whole to release nitrogen as NH_4^+ (ammonium) or NO_3^- (nitrate). It has also been proven that plants can take up entire complex molecules through root cell membranes by a process called pinocytosis[96].

This process enables plants to utilize amino acids, amino sugars, nucleic acids, and other nitrogen compounds without their initial breakdown to the most simple reduced or oxidized nitrogen forms.

When nitrogen from any source is added to the soil, the first response is for soil bacteria and fungi to consume it. Nitrogen is usually in short supply, so bacteria and fungi attack it to build protoplasm, even outcompeting plant roots for nourishment. As the saying goes, "Microbes eat at the table first."

It used to be thought that soil organisms first consumed the nitrogen and immobilized it into their protoplasm. At microbial death this nitrogen became a part of soil humus, which subsequently over the months and years was gradually released by bacterial and fungal enzymatic attack. This scenario is revealed in the nitrogen cycle diagram. While there is truth within this scheme, it is overly simplistic and does not account for the countless interactions with "grazers" of those microbes. In fairly recent studies involving the so-called "soil foodweb", it has been revealed that bacteria and fungi immobilize nitrogen in their biomass as quickly as it is added to the soil[97]. Nearly all available soil nutrients as well are thought to be contained within the microscopic bodies of these one-celled organisms. This immobilization step can be envisioned in Figure 3-19[98].

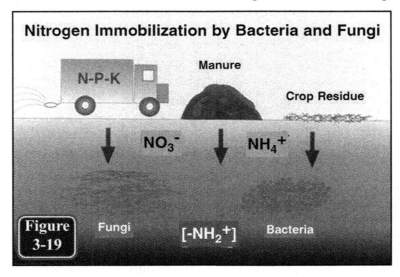

Figure 3-19

Nitrogen Immobilization by Bacteria and Fungi

N-P-K   Manure   Crop Residue

NO_3^- NH_4^+

Fungi $[-NH_2^+]$ Bacteria

Nitrogen Release by Soil Organisms

Once immobilized, a properly structured and oxygenated soil that contains a complexity of aerobic (oxygen-loving) organisms will begin releasing NH_4^+, NO_3^-, and various other nitrogen compounds through the excreta of protozoa, nematodes, microarthropods, and earthworms. Protozoan and bacterial feeding nematodes consume bacteria, and microarthropods and fungal feeding nematodes consume fungi. The excreta of these larger microbes can produce as much as 80% of the plant-available nitrogen that occurs in the soil[99] ... a phenomenal fact that is seldom appreciated. This release process is summarized in Figure 3-20[100].

Although the release of nitrogen and other nutrients in this scenario involves the consumption of one organism by another, it is really little different than a cow grazing on pasture grasses, a symbiotic system whereby the cow is benefitted by acquiring energy, minerals, and vitamins from the grass and the grass serves the cow from nutrients in the soil. The grass will regrow ... and in fact is stimulated to grow even more from Epidermal

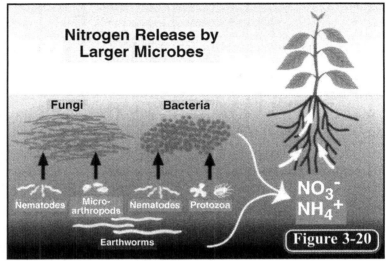

Growth Factor (EGF) found in the saliva of the cow[101]. The consumption of grasses and herbaceous, green plants was prescribed by the Creator of the ecosphere, who specified,

> "And to every beast of the earth, and to every fowl of the air, and to every thing that creeps upon the earth, wherein there is life, I have given every green herb for food..."[102].

Thus, it is a highly mutualistic system, for it was created **good** by the One who made the heavens and the earth.

Earthworms operate within the same mileau as the microbes. They ingest soil — especially the more highly organic bits of it — and decomposing organic residues, plus fungi, bacteria, and other organisms that are embedded within the soil matrix. On fields receiving regular and ample manure applications, numbers may be as high as 1,000,000 per acre, whereas on non-manured land the number may be closer to 10,000 per acre[103]. The living tissue of earthworms can total over 1,000 lb/acre, about the weight of a cow.

Earthworm castings may amount to as much as 15 tons/acre (dry), meaning that an acre-furrow slice would be totally processed in 60 to 70 years. During digestion, the mineral and organic materials are attacked by enzymes and are ground in a gizzard, leaving the castings enriched with nitrogen and available nutrients. To illustrate the power of earthworms to enrich soils over a short period of time, note the following data from an experiment in Ohio where coal mine spoilbanks were either populated or not populated with earthworms. After 175 days the soil was analyzed, showing the great increase in available nutrients that earthworms can generate (Figure 3-21)[104].

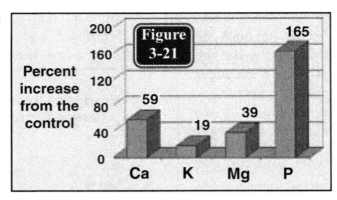

Figure 3-21

The Carbon-Nitrogen Ratio

Associated with microbial activity throughout the soil mass is the carbon-nitrogen ratio. This ratio of percentage of carbon to nitrogen in the soil is usually between 10 and 12 to 1, and varies little in soils within a region of similar soils. This relationship between carbon and nitrogen is important to understand because it relates directly to the availability of nitrogen to plants. Since soil microbes "eat at the first table", if there is a limited amount of nitrogen in the soil compared to the carbon of added organic materials, then the microbes will hold and recycle amongst themselves the nitrogen, tenaciously hanging on to it until the high-carbon compounds are totally reduced to humic substances. Once reduced, then nitrogen will again be made available to plant roots.

The microbes are not being stingy and selfish; they are just doing the job they were programmed to do. Usually the tie-up of nitrogen is just local in the soil environment where occasional bits of highly carbonaceous plant residues have been deposited. However, if an entire field area is loaded with fresh high-carbon residues such as corn stalks — especially if they are mixed into the top few inches of soil — then a massive tie-up of nitrogen can be anticipated for some time. Plants growing in such a situation will likely suffer from nitrogen deficiency, and possibly other nutrient deficiencies as well since nitrogen availability is so highly related to the uptake of the complete gamut of essential nutrients.

Manure and legume residues may have carbon-nitrogen ratios of 20 or 30:1, whereas straw and corn stalks may reach as high as 90:1[105]. Microbial bodies normally have a carbon-nitrogen ratio of 4:1 to 9:1, the bacteria generally being lower since they have more protein than fungi and other microbes. It can thus be understood why bacteria and fungi crave nitrogen, and will retain whatever they find in the soil around them until the residues upon which they are feeding have been reduced to around 10 or 12:1. This shows the wisdom of applying high-carbon residues to the soil surface — above the root zone — so the zone of nitrogen starvation will extend very little into the soil. Note differences between surface and subsurface applications of residues on nitrogen availability in Figure 3-22[106].

When organic residues of a high C:N are mixed with the soil, they first

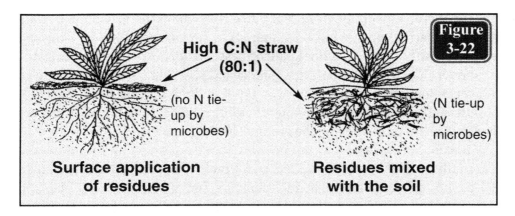

High C:N straw (80:1)

(no N tie-up by microbes)

(N tie-up by microbes)

Surface application of residues

Residues mixed with the soil

Figure 3-22

depress the nitrate level, during which time organisms break down the compounds and generate humic substances. The following graph (Figure 3-23) shows this principle, and how the resultant soil nitrate level is generally higher than at the beginning[107].

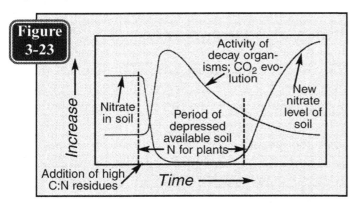

Figure 3-23

Activity of decay organisms; CO_2 evolution

Nitrate in soil

Period of depressed available soil N for plants

New nitrate level of soil

Increase

Addition of high C:N residues

Time

More than anything, this graph shows the inadvisability of incorporating high C:N residues into the soil. Because the microorganisms are working hard to decompose the residues and create humus — having a much lower C:N than the residues — they tie up the available nitrogen within their own bodies and deny the roots their needed quota for optimum growth. Such a scenario would not develop if the high C:N materials were left on the soil surface.

Contributions from the Air to Soils

We seldom appreciate the continual rain of nutrients upon the soil to gently but significantly nourish the soil surface, and leaves that might catch them. We have no trouble seeing a fraction of an inch to several feet of volcanic ash deposited to the lee of Mount St. Helens after its infamous eruption in 1980. The dust cloud of Mount Pinatubo in the Philippines in 1991 circled the Northern Hemisphere, tainting sunsets for months. This material gradually rained down on the earth over many months and years.

There are also significant amounts of wind-blown dust that periodically are wafted skyward, especially from arid areas like the Sahara, Gobi, or Kalahari Deserts, and carried downwind for hundreds or thousands of miles. During centuries past, a silty-clayey material called **loess** was deposited to the leeward side of great river flood plains or deserts. Illinois and Iowa are covered to a large extent by a layer of loess that varies from over 100 feet thick in places near the Missouri and Mississippi Rivers on the west, to several feet thick along the western portions of those states, to only a few inches in the eastern parts[108].

Recent discoveries of microcomets striking the earth's atmosphere — about a million times a day — by Dr. Louis Frank at the University of Iowa[109], has added to the understanding that micrometeors and space dust may play a more important role in soil fertility than previously thought. In millennia past it is likely that collisions with sizable heavenly bodies contributed to our topsoils as they collided with the atmosphere and deposited minerals.

Rhizospheric Soil
Definitions and Importance

Soil along the root surface is very different from the bulk soil mass. Under the influence of the plant, it is very energy-rich and supports microbial populations that reach a billion cells/cc of soil, populations 10 to 100 times larger than in soil only a few mm away[110].

On and near the root surface an incredible level of activity occurs, both chemical and biological. This activity is triggered by the programmed extension and metabolic activities of the root, and occurs in the primary zone within which the plant obtains its vital nourishment. It is also the zone of intensive mutualism between the plant and its carefully cultivated, well-fed "garden" of microorganisms, the "camp of friendship" amongst symbiotic compatriots but also the battlefield of archrivals ... the good and the bad. Thus, when we speak of the rhizosphere we are actually speaking about roots and their activities. While at any one time the rhizosphere comprises only 1% of the soil mass, yet during the lifetime of a plant the continual initiation of new roots leads to much more than 1% of the soil being explored by roots.

According to K.F. Baker and R.J. Cook, the rhizosphere is important to many disciplines:

"To the soil microbiologist, the rhizosphere is the narrow soil zone surrounding living plant roots which contain root exudates, sloughed root remains, and large populations of microorganisms of various nutritional groupings. To the plant physiologist it is the zone of ion uptake and exchange, of oxygen and carbon dioxide exchange, and of the mucigel matrix. To the soil physicist it is the zone of minimal porosity, of water diffusion and uptake, and of water-potential gradients. To the plant pathologist it is the zone where root pathogens are stimulated by root exudates, and where they swarm, grow ectotrophically, or form infection structures prior to pathogenesis"[111].

Rhizospheric soil extends to little more than 1 mm from the root surface, but effects are noted at least 5 mm away since microbes can be attracted to the root from this distance[112]. The extent of rhizospheric soil to the total mass is never very large, however. For instance, a 16-week-old winter rye plant produces 13,000,000 root axes and laterals with a total length of more than 50 km (31 miles) and a surface area greater than 200 m². In spite of this, root hairs are very fine and will occupy only about 1% of the total soil volume in the upper 15 cm (6 inches) of soil. Roots seldom occupy more than 5% of the soil volume[113].

Growth of Root Systems

Seeds of many plants produce seedlings whose radical develops into a tap-rooted system. The root system of *long-lived taprooted plants* elongates downward, developing lateral branching in two planes at right angles to each other, and at a right angle to the primary tap root. Only the taproot is strongly oriented downward (geotropically)[114].

During the first 24 hours of elongation, all roots are smooth and contain only cells undergoing division and enlargement. Then, during the next 24 hours the root cap, containing cells undergoing division and elongation, is forced forward. An electrical current is generated along the root, with a negative potential induced at the root tip making this area relatively alkaline, since protons (H⁺) and other cations are taken up from the soil[115]. The electronegativity of the negatively-charged colloidal clay and organic matter is exceeded by the electonegativity of the root tip itself, allowing for an easier uptake of cations during exchange reactions (Figure 3-24)[116].

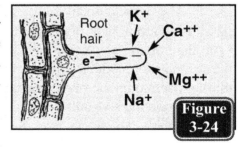

Figure 3-24

On tissues older than 20 hours, cells begin to differentiate more intensively and root hairs form. In roots two to five days old, lateral roots emerge in two planes at right angles to each other. A few days later, tertiary branching similarly develops from the laterally oriented secondary branching. Cortical cells in sections of roots about a week old turn brown and deteriorate, but stele cells continue to transport substances and develop branch roots. Basic root structure is shown

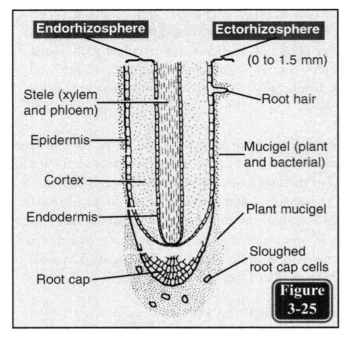

Endorhizosphere

Stele (xylem and phloem)

Epidermis

Cortex

Endodermis

Root cap

Ectorhizosphere

(0 to 1.5 mm)

Root hair

Mucigel (plant and bacterial)

Plant mucigel

Sloughed root cap cells

Figure 3-25

in Figure 3-25, with the ectorhizosphere and endorhizosphere depicted[117].

Root development is highly dynamic for about three weeks, until the initiation of fruiting. At this time root tips can thrust ahead 75mm (three inches) in 24 hours. Thousands of new roots and fine rootlets are initiated every day. Most elongate for a few days and then deteriorate, only to be replaced by others permeating nearby soil. During this period of rapid growth an area the size of a postage stamp is penetrated by roots as many as 20 times. Root and top growth cease at fruiting, but initiate anew in perennials after fruit maturation[118].

Short-lived taprooted plants, such as grasses, produce an initial taproot at seed germination, but the tap root gradually gives way to other roots that initiate from primordia located near nodes at the base of the developing stalk. As the taproot system dies from a few days to two or three weeks after emergence, the nodal root system elaborates on the below-ground-level part of each stalk, or from leaf and branch scars on buried shoots. Each stalk developing from the initial stalk develops its own nodal root system and becomes independent[119].

The typical nodal root system sends a partial ring of thin roots downward from the bottom stalk node, forming a conical surface at an angle of about 50º from the horizontal. A few days later, a second flush of thin roots develops from the second stalk node, following the same pattern. Other nodes form similar roots until a more complete ring of nodal roots is formed at about the fifth node. Long-lived taprooted plant roots can elongate three inches per day[120]. Nodal roots develop secondary laterals in two planes much like the taproot, and additional branching follows the same pattern. Dominant roots can reach several meters in length and penetrate to two meters deep in less than a month. However, nodal roots tend to develop a denser cover of root hairs than do roots of long-lived taprooted crops (Figure 3-26)[121].

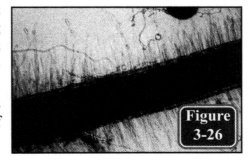

Figure 3-26

Roots must continually explore new soil for less soluble nutrients such as Ca, Mg, Zn, Fe, Cu, and Mn, so roots are constantly being formed during a plant's life. A root tip may produce 18,000 new cells each day[122]! When nearby soil has been mined of its insoluble nutrients, the cortex tissues, which are responsible for the exudation of mucigel, break down. Bacteria that abound on this mucigel then are replaced by slower growing fungi and actinomycetes that feed on more complex carbohydrates.

Changes in Rhizosphere Soil Versus the Bulk Soil

Some overall changes that occur at the root-soil boundary are discussed below[123].

1. **The structure of the soil along the root surface is altered**. Soil at the root surface is compressed, with an increase in micropores and fewer voids than in the bulk soil.
2. **A high negative water potential is created**. As water transpires from leaves the demand for soil water increases. Rhizosphere organisms may suffer from dehydration during periods of drought and function less efficiently, especially when roots are widely spaced or the soil transmits water slowly ... especially with compaction.
3. **Salts may accumulate along the root**. As the mass flow of water carries Na, K, and other ions to the root surface, some are absorbed selectively through active metabolic processes using carriers. Those ions excluded will accumulate, or even precipitate and form crystals. The osmatic potential will then be raised and may inhibit microbial activity.
4. **The pH may be more than two pH points different than the bulk soil**. As anions like NO_3^- are taken up, $H_2PO_4^-$ is secreted, or when NH_4^+ is absorbed, H^+ is secreted.
5. **Oxygen and CO_2 levels change along the root due to root and microbial respiration.**
6. **Root cells may release ethylene, terpenes, or other compounds that inhibit microbial growth.**
7. **Organic matter of the rhizosphere soil is altered**. The root cap secretes a carbohydrate-rich gel that causes the outermost cap cells to slough off, easing root passage through the soil. Root hairs likewise secrete mucilage at their tips as they move into soil crevices, and some epidermal cells rupture and spill their contents into the adjoining soil. Exudates from the root include both simple compounds such as sugars, organic acids, and amino acids and more complex substances such as plant hormones and vitamins. Up to 25% or more

of the plant's total photosynthetic energy may end up being exuded into the rhizosphere. A wheat crop may deposit more carbohydrates in the soil during growth than it deposits in the grain.

8. **Weathering of mineral particles is accelerated**. Chelating agents and organic acids from rhizosphere organisms or the roots themselves increase the rate of breakdown of soil minerals and help release nutrients.

Although there are huge numbers of bacteria in the rhizosphere, only about 7 to 15% of the root surface is occupied by microbes[124]. As the roots age, the amount of exudation decreases. However, dead cells provide a food source for many organisms, especially fungi and actinomycetes that feast on cell wall lignin and cellulose.

Organisms and Processes in Rhizosphere Soil[29.]

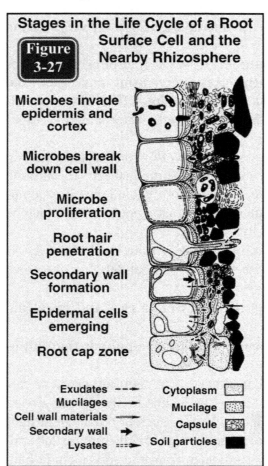

Stages in the Life Cycle of a Root Surface Cell and the Nearby Rhizosphere

Figure 3-27

Microbes invade epidermis and cortex

Microbes break down cell wall

Microbe proliferation

Root hair penetration

Secondary wall formation

Epidermal cells emerging

Root cap zone

Exudates `---▸`	Cytoplasm ☐
Mucilages `——▸`	Mucilage ▨
Cell wall materials `——▸`	Capsule ▩
Secondary wall `➤`	Soil particles ■
Lysates `====▸`	

Along with this introduction to the activities within the rhizosphere, let us examine more clearly the mutualism that exists between the plant, the rhizospheric soil, and the soil organisms that inhabit the rhizosphere. Figure 3-27 pictures the several stages in root cell development and nearby rhizosphere soil interaction as the cell is first produced by division at the root cap, and then matures and finally disintegrates as the cell ages[125]. The overall activity of the root is summarized in the Figure 3-28[126].

The plant expends a great amount of its energy, captured from sunlight and transferred to carbon compounds, in building root cells excreting high-energy mucigel into the soil alongside the root surface. About 12 to 40% of fixed energy — in some cases even more — is expended on growing this microbial garden along the root surface, much to the mutual benefit of both the plant and the organisms. Cases have been documented where 70 to 85% of the material transported to the root has been deposited in

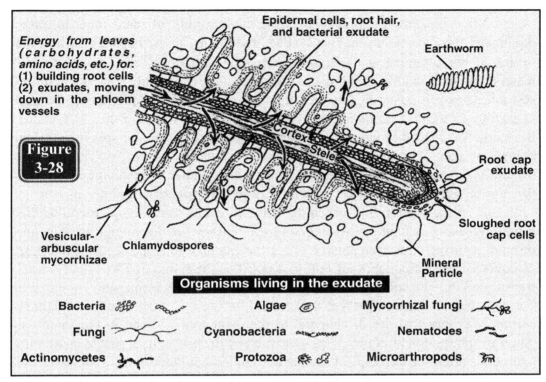

Figure 3-28

Energy from leaves (*carbohydrates, amino acids, etc.*) for: (1) building root cells (2) exudates, moving down in the phloem vessels

Epidermal cells, root hair, and bacterial exudate

Earthworm

Cortex Stele

Root cap exudate

Sloughed root cap cells

Mineral Particle

Vesicular-arbuscular mycorrhizae

Chlamydospores

Organisms living in the exudate

Bacteria   Algae   Mycorrhizal fungi

Fungi   Cyanobacteria   Nematodes

Actinomycetes   Protozoa   Microarthropods

---

the rhizosphere[127]! The plant root together with the adjacent soil create the environment favorable for the needed organisms. The organisms obviously enjoy the luxury of a highly nutritious food source which allows the populations of various species to explode into the billions per gram of soil directly along the root surface. In Table 3-3 is given a selection of some components of this mucigel exudate[128].

Amino acids (asparagine, methionine, adenine, serine, aspartate, valine, glutamate, leucine, lysine, tryptophan, tyrosine, glutamine, phenylalanine, histidine, arginine, alanine, glycine, proline)		
Vitamins	**Indole**	**Tartaric acid**
Sugars	**Salicylic acid**	**Oxalic acid**
Tannins	**Purines**	**Malic acid**
Alkaloids	**Pyrimidines**	**Citric acid**
Phosphatides	**Nucleic acids**	**Scopoletin**

Table 3-3

It is thought that the plant can even adjust the composition of its exudate so that various plant nutritional needs, communicated through the means of these exuded compounds, can be met through the by-products of the organisms thus encouraged to proliferate[129]. Thus, plants can be thought of as having intelligence of a sort, able to adjust their responses in the face of nutritional needs and other environmental stresses. It is known, for instance, that deficiencies of Fe and Mn will trigger some plants to secrete more chelating agents in the mucigel to complex these ions[130].

What is seldom appreciated is the likely ability of plant roots to attract organisms through electromagnetic (EM) radiations, or perhaps by some other means of non-attenuating radiation. Very long radiations (such as 10^6 to 10^{10} meters) can pass readily through soil and rock, so it is possible that root cells and soil microbes can communicate remotely, the root alerting a microbe of the need to migrate (if it can) by cilia or other means to populate the root zone. As Callahan has said, "Every cell of every individual tissue of any particular species — plant, animal, or man — is characterized by its own oscillatory shock[131]." The same can be said for every microbe in the soil. A constant chatter of language must therefore ensue within the plant-soil system, if only we could hear and decipher it.

It is likely possible for healthy roots to excite microbes to travel to their proximity using certain wavelengths of EM energy, or discourage certain ones through emitting other frequencies. Take the example of weed seeds. It has been discovered that a frequency of 600 to 4,000 Hz, the range of EM "anaesthesia", transmitted into the soil will cause weed seeds to store this information and remain dormant, even if the frequency is transmitted for only a few seconds[132]. Might not certain microbes that are detrimental to the plant root respond in the same way when an "anaesthetic frequency" is generated by the root cells ... just as those same pathogens might be attracted by frequencies of sick or deteriorating roots ... thereby mobilizing "nature's cleanup crew" into action? Such frequency effects have

Figure
3-29

been noted in research with field crops and their predators, such as corn and earworm moths. In the infra red to ultraviolet spectra (about 10^{-3} to 10^{-7} meters, or 10^{12} to 10^{16} Hz) insect pheromones operate. Chemical messengers from females interact with light wavelengths to attract males, which receive these frequencies with their antennae, body hairs, or other organelles[133]. See a picture of a moth antenna in Figure 3-29[134]. Moths, beetles, and many other insects are insatiably attracted to fluorescent and filament bulbs, which give off infra red radiation since their mating instincts are excited by the light's frequency, which mimics the pheromone frequency.

There are other types of radiation that could also be involved with attracting or discouraging microbe movement to and from roots. These radiations may be of a type that does not attenuate (grow weaker with distance), and could be associated with Kirlian emissions, which are known to be emitted by all living and nonliving things.

While the plant excretes exudate to feed the organisms along the root surfaces, the proliferating organisms produce by-products that are taken up by the roots, completing the symbiotic cycle. Note this process in Figure 3-30[135].

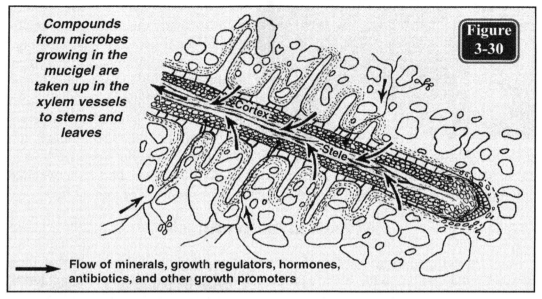

Compounds from microbes growing in the mucigel are taken up in the xylem vessels to stems and leaves

Figure 3-30

→ **Flow of minerals, growth regulators, hormones, antibiotics, and other growth promoters**

    The mutualistic interaction between the roots and soil can then be pictured as a continuous cycle, called "The Symbiotic Cycle". The plant feeds the rhizosphere organisms and the organisms feed the plant in a cyclical, simultaneous process. Both processes can occur together in the cells of the root because there are separate channels for solution flow within and between cells of the root epidermis and cortex[136]. The phloem vessels, moving photosynthate down into the roots, are separate from the xylem vessels that return water and nutrients up into the stems and leaves (See Figure 3-31)[137]. The structure and operation of these vessels are described in an earlier section. While xylem and phloem vessel elements move water and solutes on a large scale, *microtubules* transport solutes in their fine tubular network between cells of all types. See micro-

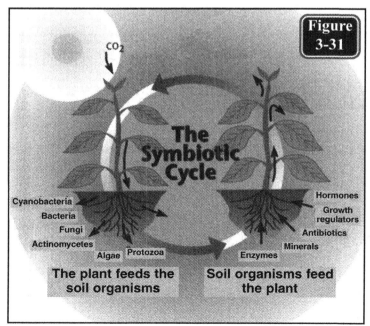

Figure 3-31

CO_2

The Symbiotic Cycle

Cyanobacteria
Bacteria
Fungi
Actinomycetes
Algae Protozoa

Hormones
Growth regulators
Antibiotics
Minerals
Enzymes

The plant feeds the soil organisms

Soil organisms feed the plant

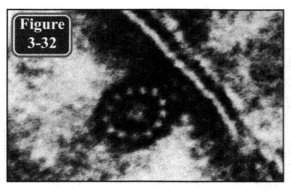

Figure 3-32

tubules in Figure 3-32[138].

Notice that the array of substances which the microbes produce includes many that are highly beneficial to plant growth: vitamins, hormones, growth regulators, enzymes, antibiotics, organic acids, and mineral elements. The details of chemical and biological mutualistic interactions within the rhizosphere are absolutely astounding on close examination. Assuredly only a small portion of these mutualistic interactions, so beneficial to both the plant and its associated microorganisms, have been unveiled to date. Only a few are presented here.

Mycorrhizal Symbiosis

Specialized fungi are found in most soils throughout the world that form intimate associations with plant root cortex cells. These fungi are called **mycorrhizae**, meaning "fungus root". They are so important that some researchers term the entire rhizosphere plus the zone to which mycorrhizal hyphae extend (to over 3 inches in some cases) as the **mycorrhizosphere**. They come in several forms, the major groupings of which are listed below[139].

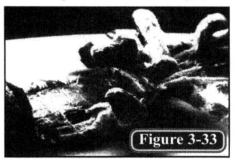

Figure 3-33

Ectomycorrhizae. These fungi form symbiotic associations primarily with pine trees and associated species. They form a structure called a Hartig net that ensheaths the root cap, forming an intimate contact with these cells (Figure 3-33)[140].

Endomycorrhizae (also called *vesicular-arbuscular mycorrhizae*). This most common type of mycorrhizae (Figure 3-34)[141] inhabits the roots of most important food and forage crops, as

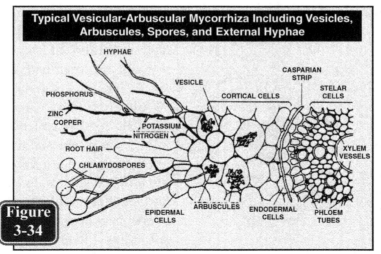

Typical Vesicular-Arbuscular Mycorrhiza Including Vesicles, Arbuscules, Spores, and External Hyphae

HYPHAE
CASPARIAN STRIP
VESICLE
PHOSPHORUS
CORTICAL CELLS
STELAR CELLS
ZINC COPPER
POTASSIUM
NITROGEN
ROOT HAIR
XYLEM VESSELS
CHLAMYDOSPORES
ARBUSCULES
EPIDERMAL CELLS
ENDODERMAL CELLS
PHLOEM TUBES

Figure 3-34

well as other plant and tree species. They form intimate associations with root cortical cells, creating a placenta-like "arbuscule" within the cells but not actually breaking the cell membranes. Across this greatly expanded feeding surface area photosynthate from the plant can be fed into the fungus, and nutrients from the fungus can be shunted into the root cell and on up the xylem stream to the leaves. They are labeled "vesicular" because these mycorrhizae generate vesicles in the soil outside the root, while creating arbuscules within root cells.

Ectendomycorrhiza. These fungi form symbiotic associations that are somewhere between ectomycorrhizae and vesicular-arbuscular mycorrhizae. A Hartig net of some degree is formed, but there is also a high degree of intracellular penetration. They are most commonly found with pine roots.

Arbutoid mycorrhiza. This fungal type forms a Hartig net plus mildly infects intercellular spaces, and may form a sparse surface layer of mycelia.

Monotropoid mycorrhiza. With this type of fungi the roots are totally covered by a fungal sheath through which nutrients must pass.

Ericoid mycorrhiza. Specialized "haustoria" are formed as well as a sheath and Hartig net, plus extensive colonization of cortex cells in the roots of Ericaceae trees and shrubs.

Orchid mycorrhiza. These mycorrhiza infect only orchid roots, but are important to their survival.

In each case the fungi form an intimate association with the plant root, some types on the exterior of the root cap ("ecto-") and other types within the cortical cells. They extract nutrients from the root for their major food source. In

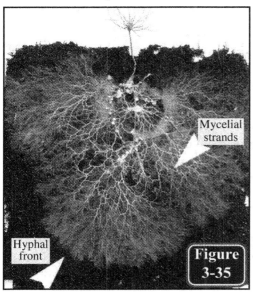

Mycelial strands

Hyphal front

Figure 3-35

every case, the fungal hyphae extending out into the soil — up to 8 cm (3.15 in)[142] — take up soil nutrients or synthesize nutrients themselves, using a portion to grow more hyphae and a portion to return to the root where the nutrients are taken up by the plant for accelerated growth. These fungi essentially function as supplemental roots for the plant, greatly expanding the volume of soil which can be explored by the root system. In pine seedlings, the ratio of the absorptive mycelium to the root can be as high as 100,000:1[143]! Notice in Figure 3-35 the tremendous ectomycorrhizal growth with the pine seedling[144].

The importance of mycorrhiza to plant growth has been demonstrated many times with a wide variety of plants. Note in Figure 3-36 the stimulated

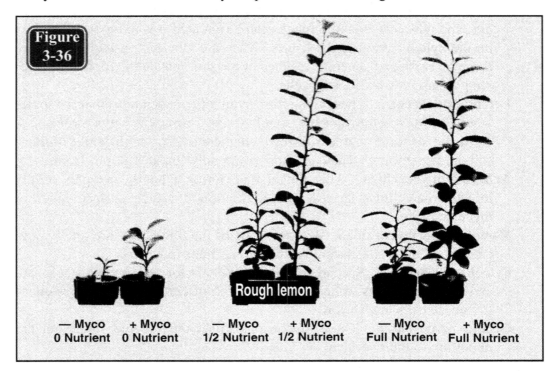

Figure 3-36

Rough lemon

— Myco / + Myco / — Myco / + Myco / — Myco / + Myco
0 Nutrient / 0 Nutrient / 1/2 Nutrient / 1/2 Nutrient / Full Nutrient / Full Nutrient

growth of rough lemon stock with VA mycorrhiza at zero, 50%, and 100% nutrients[145]. This study shows how there is an apparent improvement in nutrient uptake efficiency with mycorrhiza, one of its many benefits to plants.

To illustrate the many benefits vesicular-arbuscular mycorrhiza impart to plants — truly mutualistic effects — note the following points[146]:

(1) **Markedly increase crop growth and development**
    Citrus — up to 1600%
    Soybeans — up to 122%
    Peaches — 80%
    Pine — 323%

(2) **Stimulate the absorption of soil phosphorus, zinc, iron, calcium, copper, magnesium, and manganese**. Because these elements are immobile in the soil — i.e., firmly attached to the colloidal exchange complex — they need either direct root or fungal hypha contact to extract them from the soil matrix. The greatly expanded feeding volume due to mycorrhiza proliferation allows for the interception of many more of the immobile elements, the hyphae extending past zones of nutrient depletion. Some have calculated that about 50 cm of mycorrhizal hyphae are needed per

cm of root to account for all of the phosphorus taken up by mycorrhizal plants. Plant roots having a low surface to volume ratio, such as fleshy roots like onions and carrots having few root hairs, benefit the most from mycorrhizal fungi. Mobile nutrients like nitrate and potassium are rarely benefited.

(3) **Enhance water transport to roots**. Mycorrhizal fungi can endure much dryer soil conditions than can plant roots. This may be because of improved nutrient status of the plant due to the fungus, or to a direct effect of the fungus being able to extract water from a larger soil volume under higher tensions than roots.

(4) **Improve the tolerance of plants to toxic effects of salts and certain other nutrients at toxic levels**, such as manganese and aluminum.

(5) **Increase nodulation by symbiotic nitrogen-fixing bacteria like Rhizobium**, and resulting nitrogen fixation and dry matter production[147].

(6) **Stimulate other rhizosphere organisms**.

(7) **Provide resistance to plant diseases**. Diseases are not eliminated, but symptoms are reduced in severity. This could be due to mechanical protection from a fungal mantle, better plant nutrition, production of antibiotics, competition for infection sites, the formation of phytoalexins, alteration of root exudates, or filling the root cells with arbuscules to prevent nematode invasion.

Legume Symbiotic Nitrogen Fixation

This form of mutualism takes several forms, a major one involving specialized nodules on root surfaces within which Rhizobium bacteria utilize plant nutrients to reduce atmospheric nitrogen into plant-usable reduced nitrogen. Nitrogen is often limiting to plant growth, so its provision to plants is critical in the face of an atmosphere that is comprised of about 79% dinitrogen gas.

Rhizobium symbiosis in legumes.

The legume-Rhizobium bacteria symbiosis is one of the most well-understood and utilized mutualistic interactions between plants and soil organisms. Farmers around the world apply these bacteria to legume seeds — each legume species having unique strains designed for them — using a slurry or powder, to inoculate the plants with an efficient nitrogen fixing strain. Subspecies of bacteria can vary considerably in their ability to fix nitrogen. What is less appreciated is the fact that Rhizobia will associate with non-legume roots as well, although nitrogen fixation is thought to be less and without nodulation

To draw Rhizobium bacteria to the root, it is presumed that the root exudates of the legumes roots send "messenger molecules", or perhaps frequencies, to call in the required and mobile microbes. This is a case of the plant intelligently signaling a need, and the compliant microorganisms seeking out the opportunity to serve the host plant. Of course, the microbe stands to benefit from the calling as well, since the plant provides a home for it. One noted author claimed, "It is likely that when the final chapter is written on specificity, both lectins and extracellular polysaccharides will have been shown to have critical roles in recognition and attachment of the bacteria to host plants[148]".

Once the legume root has sent out its signal and the Rhizobia have moved to the root hair surface, they become attached by a lectin glycoprotein produced by the plant. If the Rhizobia are compatible with the root, the bacteria exude an acidic polysaccharide that facilitates the attachment. Just before invasion of the root by the bacteria the root curls, and a tube enters the root from the bacterial cell. Both the bacterium and the host root then generate multiple cells that produce the nodule (Figure 3-37)[149]. Within the nodule (Figure 3-38)[150] the

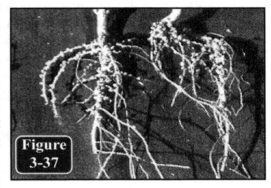

Figure 3-37

Rhizobia change from mobile to sedentary in nature and begin producing nitrogenase, a protein rich in iron and molybdenum that enables the bacteria to fix atmospheric nitrogen[151]. Nodules also produce a molecule virtually identical to hemoglobin that binds oxygen (O_2), which would poison the anaerobic action of the fixation reaction. Thus, healthy nodules are pink when cut open, revealing an active nitrogen fixing machinery[152]. A brief review of the metabolic processes that occur within the nodules is shown in Figure 3-39[153].

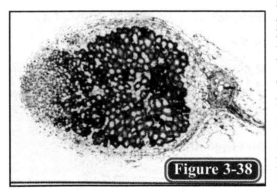

Figure 3-38

The plant gives up fixed carbon energy to the resident rhizobium bacteria, but in return gains much-needed nitrogen. Nitrogen-containing compounds make up about 25% of a plant's dry weight, although it comprises only 1 to 2% of total plant dry weight[154]. It is noteworthy that high to moderate levels of nitrate in the soil slow or stop the nitrogen fixation process of rhizobium bacteria, presumably through a plant feedback mechanism that "tells" the bacteria to stop producing nutrients already present in ample supply.

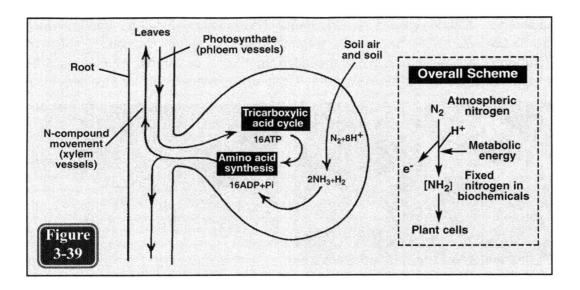

Figure 3-39

Frankia symbiosis in woody perennials

The genus *Frankia* represents filamentous actinomycetes, which are able to nodulate the roots of a wide variety of woody non-leguminous trees and shrubs, especially alders and certain tropical pines. The fixed nitrogen produced by this microorganism is highly beneficial to the growth of these species of trees and shrubs. Details of the symbiosis are not as thoroughly researched as for rhizobium, but it is known that root exudates are directly implicated with the infection and development of *Frankia* nodules. A picture of

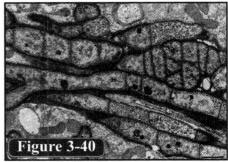

Figure 3-40

From Lynch, *The Rhizosphere*, ©1990 Wiley Interscience. Reprinted by permission of John Wiley&Sons, Inc.

Frankia hyphae traversing an infected host cell is shown in Figure 3-40[155].

Non-Legume Symbiotic Nitrogen Fixation

Nitrogen fixing microorganisms live in the mucigel layer of plant roots just as do other bacteria and fungi, and utilize fixed carbon energy, transferred from leaves, to grow and fix atmospheric nitrogen. They also occur abundantly in rhizosphere soils where their numbers may be multiplied as much as 20-fold compared to soil outside this zone[156]. Among these microbes are a wide array of heterotrophs (those requiring a soil carbon source to live) with varied oxygen requirements, only a portion of which have yet been discovered.

The aerobes and microaerophiles include species of *Azotobacter, Beijerinckia, Azospirillum, Herbaspirillum, Nostoc,* and *Derxia.* Facultative anaerobes include species of *Klebsiella, Enterobacter, Erwinia,* and *Bacillus.* Some *Clostridium* species are also implicated in non-legume rhizosphere nitrogen fixation[157]. Typical non-symbiotic nitrogen fixers are shown in Figures 3-41 (*Azotobacter*)[158] and 3-42(*Nostoc*)[159]. Certainly cyanobacteria are also important

Figure 3-41

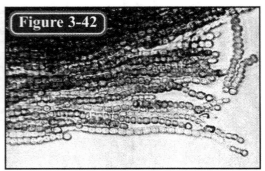

Figure 3-42

in nitrogen fixing processes within the rhizosphere, even in the absence of light, since they are capable of living as heterotrophs — using soil organic carbon sources — despite their photosynthetic capability.

The effectiveness of any of these organisms in fixing nitrogen depends on a complex array of factors: oxygen levels (directly related to soil structure), temperature, moisture content, and exudate levels and composition. The presence of nitrate, such as applied in fertilizers, shuts down or greatly inhibits nitrogen fixation. Antibiotics and other bio-regulatory compounds can also inhibit these nitrogen fixers.

The Oxygen-Ethylene Cycle

Another symbiotic association that occurs at the root surface is called the Oxygen-Ethylene Cycle, a term coined from the two major soil gases that influence the operation of rhizosphere organisms within this highly complex mineral-organism-biochemical system.

This cycle occurs directly on the root surface and extends a very small distance out into the rhizospheric soil in what are called "anaerobic microsites". It is a self-governing system whereby bacteria vigorously consume root exudates, deplete soil oxygen, and concurrently generate ethylene that is an anesthetic to the microbes. As the microbes slow down due to the anaesthetic effect of ethylene, oxygen permeates back in to the root surface and more rapid bacterial metabolism begins anew. This cyclical process is reviewed in Figure 3-43[160].

The consequences of this root surface reaction to create anaerobic microsites, however, reach far beyond merely governing the rate of microbial metabolism. Complex mutualistic nutrient release reactions occur as well.

This oxygen-ethylene cycle operates efficiently only in undisturbed forest

A Summary of Anaerobic Microsite Processes

Soil microbes (bacteria, fungi, actinomycetes, algae, etc.) grow rapidly on root exudates in the rhizosphere. These exudates originate from photosynthesis in the leaves, which are translocated into the roots.

In these microsites the microbes rapidly deplete the oxygen, allowing ethylene to be formed (nitrate inhibits the process). Iron must be reduced for ethylene to be produced in a nonbiological reaction.

Ethylene, a gas, diffuses into adjacent soil and deactivates microbes creating anaerobic microsites. Lowered oxygen demand allows oxygen to reenter microsites.

Photosynthesis and metabolism are enhanced, leading to more light energy and CO_2 trapped in carbon compounds, with greater translocation and exudation of compounds into the rhizosphere. Also, metabolites and decomposition products from microbes are available for root uptake. These include vitamins, antibiotics, enzymes, nucleic acids, regulators, and other substances. Acids produced by microbes dissolve minerals[45].

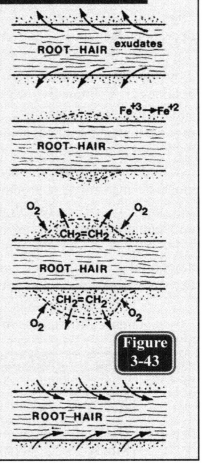

Figure 3-43

and grassland soils where there is a slow, balanced organic matter turnover. In tilled soils ethylene is seldom detected, for with tillage comes organic matter reductions, nutrient deficiencies, and structural weakening that reduces the exchange of oxygen. The operation of the microsites provides the soil yet one more way to release its minerals to the root.

Iron occurs abundantly in soil. In the oxidized [Fe^{+3}] state it occurs as minute crystals of Fe_2O_3 that have a large surface area and are highly charged. Nutrient anions such as $H_2PO_4^{-1}$, SO_4^{-2} and others are tightly bound to these surfaces and unavailable to plants, but in the anaerobic microsites the iron is reduced to Fe^{+2}, causing the crystals to break down and release anions to the soil solution. Other elements such as Ca^{+2}, Mg^{+2}, K^+, Na^+, and NH_4^+, held on the surfaces of clay particles and organic matter, are displaced into the soil solution by Fe^{+2} through mass action and are free to be absorbed by roots. Though mobile, these released nutrients become immobile once again by reacting with charged Fe_2O_3

crystals at the edges of microsites and will not leach[161].

If anaerobic microsites are discouraged, ethylene production is curtailed. High soil nitrate levels, lack of mature organic residues (ethylene precursors), or operations that encourage soil aeration discourage anaerobic microsite formation. To ensure that anaerobic microsites and ethylene are produced in sufficient amounts for successful soil management, it is important that mature organic residues are continually returned to the soil, that these residues be returned to the soil surface if possible, that minimum tillage be practiced to reduce aeration and organic matter oxidation, and that ammonium fertilizer be used if nitrogen is needed, preferable added in several small applications[162].

It has been found that nitrate inhibits the formation of anaerobic microsites and also stops ethylene production by inhibiting Fe^{+3} reduction, even if there is no iron in the Fe^{+2} form for ethylene to be produced, since only Fe^{+2} reacts with the ethylene precursor in the soil[163]. This precursor comes from old senescent plant leaves that have accumulated in the soil (and plant species differ markedly in the amount of precursor they contribute)[164]. A summary of a few of these critical reactions within the anaerobic microsites is shown in Figure 3-44[165].

For ethylene production there must be the following conditions[166]:

1. Intense rhizosphere microbial activity must create anaerobic microsites.

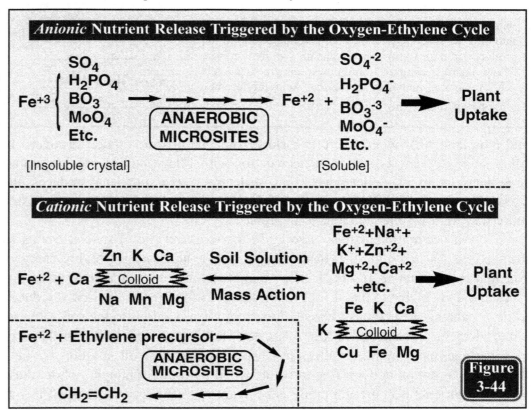

98

2. Iron must be reduced to Fe^{+2} to trigger ethylene release.

3. Nitrate in the soil must be kept to a minimum or Fe^{+2} won't be mobilized.

4. There must be enough ethylene precursor in the soil.

Once the ions are delivered to the root surface there are active uptake mechanisms that have been discussed earlier. The uptake involves active metabolic energy and carriers. Ions can move to the root via the mycorrhizae, where-

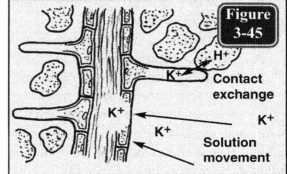

in they are carried directly to cortical cells, but they can also be brought to the root by movement in the soil solution, or by direct contact of the soil with the root (Figure 3-45)[167]. In contact exchange, an ion of hydrogen [H^+] exchanges with an ion adsorbed on the colloid.

The magnificent mutualism that occurs at the root surfaces to elicit the release of both cationic and anionic nutrients, through dissolution of the Fe^{+3} crystals to soluble Fe^{+2}, and subsequent cation exchange reactions on the soil colloids, reveals how many microbial, biochemical, and inorganic chemical participants combine in an elaborate symphony to provide food for the plant. These reactions all occur on the root surfaces, fed by the plant's exudates and microbial gardens. In Figure 3-46[168] a summary of this anaerobic microsite activity of the oxygen-ethylene cycle is shown.

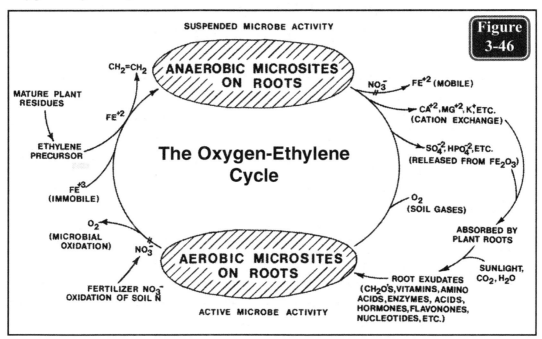

The smooth operation of the oxygen-ethylene cycle helps insure a healthy plant which is able to fend off the ravages of plant pathogens through its own defenses, which have been mobilized through root absorption of ...

a. Minerals (Ca, Mg, K, P, B, Cu, Zn, and other immobile or unavailable nutrients) which are released through dissolution by organic acids and anionic and cationic exchange mechanisms.

b. Growth regulators, especially those produced by cyanobacteria and fungi such as mycorrhizae.

c. Antibiotics, produced primarily by bacteria and actinomycetes.

d. Nitrogen compounds, produced by symbiotic nitrogen fixing bacteria such as *Rhizobium*, nonsymbiotic nitrogen fixer such as *Azotobacter, Azospirillum,* certain actinomycetes, or autotrophic organisms such as cyanobacteria.

Recent studies have revealed that ethylene generation in root systems plays a key role in plant defense mechanisms. Authors of a *Science* journal article concluded that if methods could be found to stimulate ethylene production, by the plant or by soil organisms or both — especially when plants are young — the benefits to the plant's immune system, root development, nitrogen and water uptake, resistance to pathogen and insect pressure, and overall vigor would be sizable. These benefits would tend to accrue throughout the season. Interestingly, certain biostimulants do increase the intensity of the oxygen-ethylene cycle and thus elevate the level of ethylene in the plant-soil system, producing all of the effects listed here[169].

Availability of Nutrients to Plants Is Microbially Mediated

It should be obvious at this point that soil microorganisms and some macroorganisms, such as earthworms, are highly important — in many cases essential — in making plant nutrients available. This is not to say that strictly chemical processes do not operate in making nutrients available, but rather that the chemical processes are caught up with microbial activities. For example, soil phosphate compounds release phosphorus very slowly, but the VA mycorrhiza dramatically assist the plant in collecting this and other immobile elements through its extensive hyphal network.

Many soil scientists today claim that nutrient absorption by plants is mostly chemical in nature, governed by mass flow of nutrients in soil water and by diffusion through the soil water, with direct interception by the extending root being limited except for immobile elements. Past estimates of uptake mechanisms minimized the importance of the mycorrhizae in moving immobile elements to root surfaces, or of bacteria and fungi first immobilizing nitrogen and then releasing it through the grazing of protozoa, nematodes, microarthropods, and earthworms. Thus, Table 3-4 below shows only part of the story regarding the two means by which roots receive their nutrients[170].

Chemical Mechanisms of Nutrient Uptake by Roots in a Fertile Soil, to Produce a 373 bu/acre Corn Crop				
	Plant	Amount absorbed by ...		
Nutrient	absorption	Root interception	Mass flow	Diffusion
		------------------- lb/acre -------------------		
Nitrogen	170	2	135	33
Phosphorus	36	1	2	33
Potassium	174	4	32	138
Calcium*	36	54	135	0
Magnesium*	40	14	90	0
Sulfur*	20	1	59	0

Table 3-4

*With these elements an excess is moved to the root surface, and a concentration gradient is established from the root into the soil resulting in no diffusion transfer.

These values neglect to consider the powerful role of soil organisms in the nutrient uptake process. Roots indeed intercept nutrients as they extend through

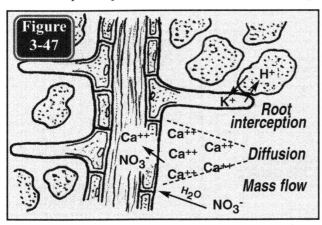

Figure 3-47

the soil, and a significant portion of some nutrients like calcium can be taken up by non-biotic root uptake mechanisms and the transpiration stream (Figure 3-47)[171]. Yet, the fact remains that the nutrients are brought into available form largely through microbial transformations. Thus, while nitrate is moved to the root in large part by mass flow, its conversion to nitrate is mostly a biochemical, microbial process.

Let us now examine each element and its transformation in Figures 3-48 through 3-54[172]. The plant-available nutrient forms are enclosed in boxes.

One phosphorus researcher has depicted microbial activity as a "wheel" that rotates in the soil, simultaneously consuming and releasing phosphorus to the soil solution[173].

It is clear throughout these nutrient transformations that soil microbes are extremely important in every case, whether release of the nutrients is from organic matter or minerals. In some cases, as with nitrogen and sulfur, the intervention by microbes is essential, while with others such as potassium and molybdenum microbes play a less prominent role. In all cases, however, the mutualism of soil

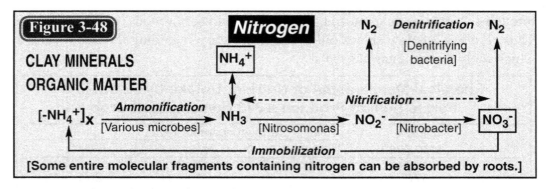

Figure 3-48

Nitrogen

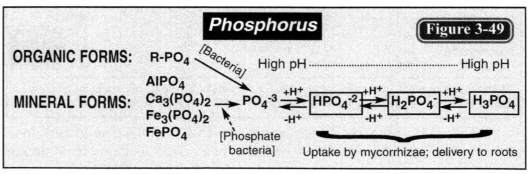

Phosphorus

Figure 3-49

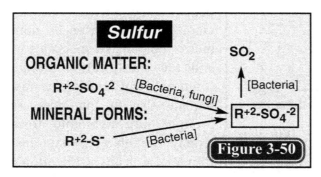

Sulfur

Figure 3-50

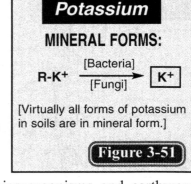

Potassium

Figure 3-51

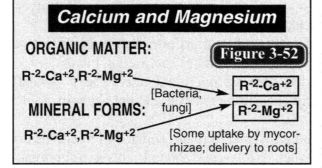

Calcium and Magnesium

Figure 3-52

microorganisms and earthworms as mediators of the nutrient supply and transformation between the soil and the root is pervasive. In normal soil environments there is an absolute requirement for bacteria, algae, fungi, protozoa, nematodes, arthropods, actinomycetes, and other organisms to transform elements into the forms that are needed by plants.

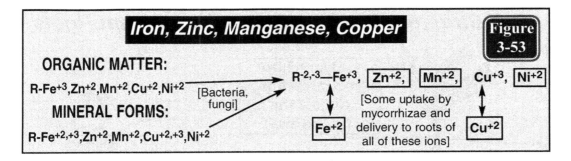

Iron, Zinc, Manganese, Copper — Figure 3-53

ORGANIC MATTER:

$R\text{-}Fe^{+3}, Zn^{+2}, Mn^{+2}, Cu^{+2}, Ni^{+2}$

MINERAL FORMS:

$R\text{-}Fe^{+2,+3}, Zn^{+2}, Mn^{+2}, Cu^{+2,+3}, Ni^{+2}$

[Bacteria, fungi] → $R^{-2,-3}\text{—}Fe^{+3},\; \boxed{Zn^{+2},}\; \boxed{Mn^{+2},}\; Cu^{+3},\; \boxed{Ni^{+2}}$

$\boxed{Fe^{+2}}$ [Some uptake by mycorrhizae and delivery to roots of all of these ions] $\boxed{Cu^{+2}}$

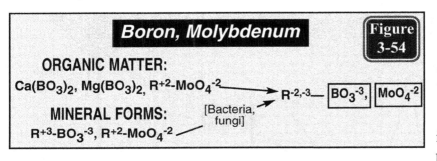

Boron, Molybdenum — Figure 3-54

ORGANIC MATTER:

$Ca(BO_3)_2,\; Mg(BO_3)_2,\; R^{+2}\text{-}MoO_4^{-2}$

MINERAL FORMS:

$R^{+3}\text{-}BO_3^{-3},\; R^{+2}\text{-}MoO_4^{-2}$

[Bacteria, fungi] → $R^{-2,-3}\text{—}\boxed{BO_3^{-3},}\; \boxed{MoO_4^{-2}}$

Soil Structure Formation

Of primary importance to a plant's overall health is an abundance of oxygen in the rhizosphere. Root cells, like all types of aerobic (oxygen-loving) cells, require at least a minimum level of this essential gas to operate efficiently. Typical compaction will reduce root growth from a maximum of three inches per day to only one-half inch per day, or 8% of growth compared to root extension in a well-structured soil[174] (see Figure 3-55). A root can penetrate an oxygen-deprived environment about one inch, due to cell elongation, but if no oxygen is found the rootlet stops growing. Cell division stops without oxygen. With less root extension comes less water and nutrient absorption, and consequently greatly reduced growth and yield.

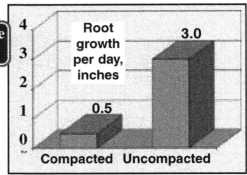

Figure 3-55

Root growth per day, inches

Compacted 0.5 Uncompacted 3.0

In an attempt to maximize its root growth potential, the plant will utilize its array of rhizosphere microorganisms to render the soil in the vicinity of the roots more strongly structured ... with numerous cleavage planes and a preponderance of macropores (larger pores) that will allow for the easy exchange of gases and the ready movement of water. The root also improves the structure so it can more easily penetrate the soil

The strategies the root uses to improve soil structure and thus improve the plant's overall health are numerous and highly symbiotic. These strategies are summarized in Figure 3-56 and are discussed below[175].

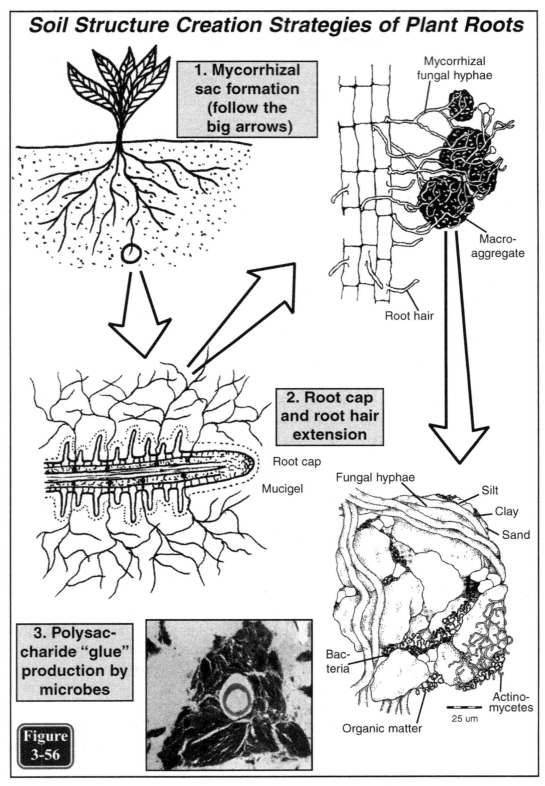

Soil Structure Creation Strategies of Plant Roots

1. Mycorrhizal sac formation (follow the big arrows)

Mycorrhizal fungal hyphae

Macro-aggregate

Root hair

2. Root cap and root hair extension

Root cap

Mucigel

Fungal hyphae

Silt

Clay

Sand

3. Polysaccharide "glue" production by microbes

Bacteria

Organic matter

Actino-mycetes

25 um

Figure 3-56

1. **Mycorrhizal fungi**, feeding on plant energy stored in carbon compounds that is moved underground in phloem vessels, proliferate from root cortex cells out into the surrounding soil, up to several centimeters out from the root. These hyphae form sac-like structures that serve to bind smaller structural units and larger soil particles such as sand grains. As shown in Figure 3-56, the mycorrhizal hyphae form the fabric within which an array of actinomycetes, bacteria, and mineral particles reside. Actinomycetes act like fungi but are smaller, and further net together soil particles, especially silt and clay. Note point 2 below regarding bacterial activity.

2. **Polysaccharides** produced by bacteria and other microorganisms in the rhizosphere are sticky, and glue together smaller soil particles such as clay and silt. Within the interiors of these "peds" the oxygen level is very low, providing for very slow oxidation and the long-term storage of organic materials. In the center of the ped in the photo of Figure 3-56, note the remains of a fungus that has contributed to synthesizing a sticky polysaccharide that holds the particles together. Much of the nitrogen in a native soil is stored long-term within the interior of the soil peds, safe for use later unless the ped is broken by the mechanical action of cultivation or some other activity.

3. **Roots and root hairs** move through the soil as they grow, forming channels which later serve as conduits for air and water movement after they have died.

In addition to these three modes of action, earthworms also move through the soil and form pores and channels that encourage air and water movement. Other soil creatures such as ants, millipedes, and mites further create pores. The total effect of these various organisms is to generate strong structural units that allow air and water movement and root growth along their cleavage surfaces. A greater number of macropores results from this community-wide mutualistic effort, affording the plant an optimum growth environment. It is the plant itself that engineers this communal effort, to optimize its existence.

The construction of a strongly aggregated soil may be likened to building a brick house. Note the following useful analogy[176].

Brick House	**Soil Structure**
Bricks. Sand, silt, and clay are bound together with straw.	*Microaggregates.* Bacteria bind together clay, sand, and silt with polysaccharides.
Walls. Mortar holds the bricks together in large units.	*Macroaggregates.* Fungal hyphae, root hairs, and roots hold together micro-aggregates.
House. The walls and roof are nailed together in various patterns.	*Larger aggregates.* Arthropods, insects, and earthworms further modify smaller aggregates into even larger structural units.

This scheme is elaborated upon in the following sequence (Figure 3-57)[177].

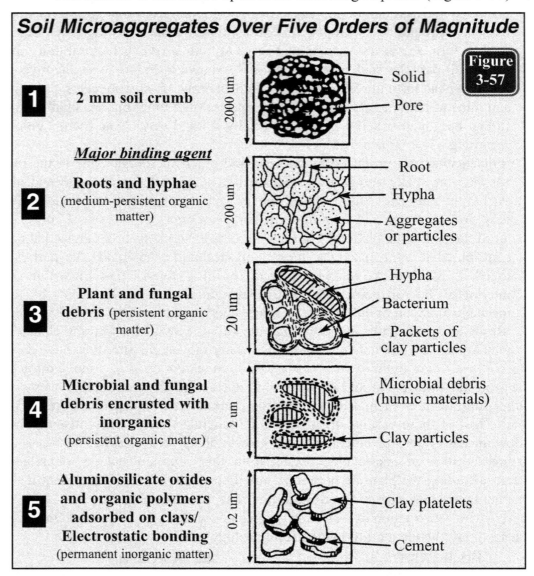

Soil Microaggregates Over Five Orders of Magnitude

1 — 2 mm soil crumb (2000 um) — Solid; Pore

Major binding agent

2 — Roots and hyphae (medium-persistent organic matter) (200 um) — Root; Hypha; Aggregates or particles

3 — Plant and fungal debris (persistent organic matter) (20 um) — Hypha; Bacterium; Packets of clay particles

4 — Microbial and fungal debris encrusted with inorganics (persistent organic matter) (2 um) — Microbial debris (humic materials); Clay particles

5 — Aluminosilicate oxides and organic polymers adsorbed on clays/ Electrostatic bonding (permanent inorganic matter) (0.2 um) — Clay platelets; Cement

Figure 3-57

Only by having a complete array of bacteria, fungi, microarthropods, insects, earthworms, and other creatures can the structure become fully stabilized, since all are required in one way or another in structure formation. The largest units are the most fragile, but are very important in that they help form and maintain the continuum of channels that extend from the surface into the subsoil.

Soil Compaction and Organisms

Soil compaction is the breaking and crushing of structural units, reducing many of the macropores to much smaller micropores that restrict oxygen and air movement, and thus greatly restrict root growth. Organisms living within these collapsed spaces are killed, including those that function within the oxygen-ethylene cycle, the mycorrhiza, and symbiotic and non-symbiotic nitrogen fixers. Thus, nutrient release mechanisms are greatly suppressed by soil compaction.

The extent to which soil compaction can occur is illustrated in Figure 3-58[178]. In this typical corn field the soil was exposed with a backhoe and the exposed surface picked away to reveal compacted layers. Note how tractor tires and implements — even the planter — contributed to these compact layers that essentially prevented the proliferation of plant roots into the soil mass, thus restricting fertilizer and water utilization.

Figure 3-58

A strong structure can withstand considerable weight bearing upon it, but a tractor or truck driving on the soil, or an implement being pulled through it, will invariably break structural units and collapse them. With a complex array of organisms in the soil which have built a strong structure, destruction of the units by heavy equipment and tillage will be less ... and recovery from compaction will generally be faster than with a weak structure, especially if there are sufficient fresh organic residues to enable a rapid recovery of killed organism populations. When the soil is wet these peds are much easier to break ... so while tillage itself tends to rupture and destroy structure, tillage when the soil is wet multiplies the destructive forces of machinery.

As compaction increases, the following changes occur[179]:

(1) The biggest organisms are most susceptible to compaction, so when they are lost the soil will switch from a fungal to a bacterial-dominated system.

(2) As compaction increases the largest of the predators of bacteria and fungi are killed. Since fungal-feeding predators are larger than bacterial-feeding predators, fungal-feeding organisms die (certain nematodes and microarthropods) ... which prevents the release of nutrients tied up in fungal biomass.

(3) As macropores are restricted even more, about the only organisms that survive are some bacteria, protozoa, and opportunistic fungi and nematodes. Pathogenic nematodes, fungi, and insects have few competitors, and roots

have difficulty extending through compacted, low-oxygen soil. Whatever bacterial growth that occurs uses up oxygen quickly, turning the system anaerobic and encouraging the toxic by-products of bacteria to be produced ... further damaging roots.

While physical mechanisms such as wetting and drying, freezing and thawing, flocculation effects of calcium and other ions, and others play a part in the formation of a strong soil structure, it is the symbiotic associations of soil microorganisms that play the major role in structure formation. This is especially true in the rhizosphere, but it is true throughout the soil mass. We owe our physical lives to the cooperative efforts of the tiny creatures at the bottom of the biotic pyramid, since they make the soil work as it does. Without them life would not be possible.

Defense Mechanisms Against Root Pathogens

Because of modern agriculture's assaults against natural laws, plant diseases have become a serious problem for farmers. Such is seldom the case when plants are grown in undisturbed, native, diverse-variety conditions. As one modern soil microbiology book states,

"Changes produced by agricultural practices tend to be varied, rapid, and ongoing, which permits little time for the establishment of the biological equilibrium. Intensive use of fertilizer, pesticides, tillage, irrigation, and other crop management practices for selected high-yielding cultivars has resulted in increased severity and incidences of plant disease. The replacement of a diverse plant community with a single genotype, or monoculture, shifts the biological equilibrium toward development of plant disease epidemics that otherwise rarely occur in undisturbed ecosystems[180]."

An axiom of nature states that a healthy plant is the best preventative against disease. At the same time, a healthy plant is the result of the countless mutualistic interactions among minerals, microorganisms, and roots that normally occur in the soil. Health is the normal state of an organism. If given the necessary tools within a natural environment, aggressive and disease-free growth will usually result.

For example, the *Gros Michel* banana was grown in tropical plantations for many years until a devastating race of *Fusarium* wilt — termed Race Four — struck the crop. The disease quickly spread to all countries and today this variety of banana cannot be cultivated conventionally. It has been replaced by banana varieties such as *Cavendish* that are resistant to this fungus.

Some of the *Gros Michel* banana plants have escaped to forested environments outside plantations. There they grow well, producing disease-free leaves and yielding good fruit bunches under the proper conditions ... sometimes only a short distance from an existing plantation. Why can they thrive disease-free in the natural state but not in the captive cultivated state? The answer lies in the effects of modern banana production on soil microbiology and plant nutrient uptake. Fertilizers and pesticides, applied continuously for many years, have greatly altered the array of microorganisms in the soil, simplifying the community and reducing the natural predators of pathogenic fungi, bacteria, and nematodes while glutting the plant with high levels of free amino acids, and creating imbalances or shortages of some elements. All of these changes in soil and plant composition have their consequences, as we shall see. Pathogenic nematode losses alone cost farmers over 12% of their annual production of major food crops[181], and a chemical bill of $25 billion (1993 data) has not stopped the carnage.

Built-In Plant Mechanisms

Plants have built within them marvelous mechanisms to defend against attack by outside invaders ... much like the human or animal body has an immune system. Plants also have an immune system. As one researcher put it, a plant's capacity to overthrow an attack by parasites is like defending a medieval castle[182].

"Extrusion from plants of chemicals toxic to pathogens is equivalent to the pouring of boiling oil on invaders by castle defenders. Preformed structures and chemicals that ward off pathogens at the plant surface are similar to the castle setup of reinforced walls, trap doors, and poisonous baits. These mechanisms constitute the first line of defense. When this defense fails, many plants synthesize toxic substances, such as apposition of callose to cell walls to fence off the invading pathogens. These defenses are similar in a way to the throwing of explosives and setting up barricades by the castle defenders to confine and minimize the damage done on the castle by invaders."

Soybean plants, for example, appear to use a cascade of defenses against an invading fungus (see Figure 3-59)[183]. Within minutes of the attack an initial blockade is formed: the plant begins cross-linking proteins in the cell walls at the invasion site, sort of like closing the slats of a venetian blind to block fungal movement. Then some of

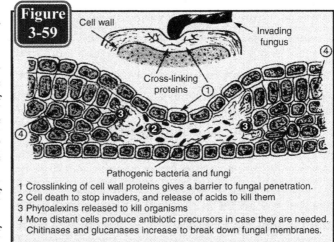

Figure 3-59

Cell wall

Invading fungus

Cross-linking proteins ①

④

Pathogenic bacteria and fungi
1 Crosslinking of cell wall proteins gives a barrier to fungal penetration.
2 Cell death to stop invaders, and release of acids to kill them
3 Phytoalexins released to kill organisms
4 More distant cells produce antibiotic precursors in case they are needed. Chitinases and glucanases increase to break down fungal membranes.

the cells around the infection die, surrounding the fungus while releasing acidic and other defensive substances that further frustrate the invader. This is called the *hypersensitive response.*

After the initial blockade to slow down fungal extension, surrounding cells are given more time to begin producing potent chemicals called *phytoalexins.* These chemicals can stunt development of the fungal mycelia, much like penicillin does to fungus colonies in a petri dish. A message is somehow sent to more distant cells to produce precursors to antibiotics in case they are needed there. Finally, the whole plant mobilizes to fight the invader. Enzymes called chitinases and glucanases, which can break down the exterior of fungi, are synthesized and disseminated throughout the plant. Parts of the plant far removed from the infection also cross-link their cell walls in case the invader penetrates further[184].

The root exudates discussed earlier may play an important role in defending against pathogens. For instance, the exudates can either attract or repel pathogenic nematodes, meaning that nematode-resistant plant roots produce compounds that are antagonistic to the pests, while non-resistant plant roots produce signals — few of which have been identified — that attract them[185]. Root exudates of certain plants like marigolds (*Tagetes erecta* L.) produce about 0.1% a-terthienyl that exhibits high nematicidal activity against several plant-pathogenic nematodes; 63% of *Meloidogyne incognita* juveniles were killed in marigold root exudate in 72 hours[186]. Nimbidin and thionemone produced by margosa (*Azadirachta indica* Juss.) roots are potent nematicides, and polyacetylenes from safflower (*Carthamus tinctorius*) roots and a-chaconine from potato sprouts also are strongly nematicidal against certain nematode species[187].

Cucumber seedlings can be inoculated against disease by using anthracnose fungus. By inoculating seedlings with the fungus to produce a mild infection, the cucumber plants are protected from severe infection for up to six weeks. Later damage is reduced by more than 98%! What is more, the immunized plants were protected from at least 12 other diseases caused not only by fungi, but also bacteria and viruses as well which infest the roots, fruit, and leaves[188].

Other researchers have similarly studied the inoculation of cotton seedlings with spider mites to increase resistance to wilt; conversely, cotton seedlings infected with the wilt fungus acquire resistance to spider mites. Injecting tobacco plants with aspirin (salicylic acid) protects them against the tobacco mosaic virus. It has recently been found that most plants, in response to necrosis of leaf tissue, produce immunity-enhancing salicylic acid. Other plants respond to insect attack by brandishing more hairs, thorns, or tougher bark, or by producing internal toxins.

These mechanisms of plant protection are all internal to the plant. Yet, the plant is anchored in the soil, so its welfare cannot be segregated from soil functions. What defense does the soil itself muster to repel parasites and other attackers, the organisms that are the opposites of beneficial, mutualistic ones already discussed? The rhizosphere arsenal of plant-protecting agents is quite awesome indeed. The organisms that are fed by the plant and its roots are highly partial to their niches; many of them will forcibly attempt to repel or destroy any invaders if given the opportunity. In some cases they may simply outgrow the invader, and more effectively compete for the carbon, nitrogen, and other nutrient sources available so the pathogen poses no threat to the crop.

These discussions will not involve the farmer's intervention into the plant-soil system to repress or annihilate a pathogen. Rather, they will discuss the built-in systems of rhizosphere organism control of pathogens. These biological control agents are more subtle and operate more slowly than toxic chemical intervention, but generally are more stable and longer lasting than chemical and cultural control measures[189].

Soil Bacteria and Fungi Attaching Pathogenic Fungi

It is well-documented that the fungus responsible for the take-all of wheat — *Gaeumannomyces graminis* var. *tritici* — is attacked by soil bacteria, in particular by the bacteria in what are called **take-all suppressive soils**. These soils are unique in that the severity of the disease becomes progressively less as the cropping season continues. In some cases the disease may not even express itself whatsoever despite being present.

Figure 3-60

Note a picture of *G. graminis* in Figure 3-60[190]. In this electron micrograph a bacterial infection has caused the collapse of the hyphae on the right, with a breach created in the wall at the arrow.

It is concluded by soil microbiologists that most soils express some degree of natural pathogen suppression[191]. This occurs generally in soils by the mass of beneficial organisms overwhelming the pathogens at a critical time in their life cycle, robbing critical nutrients from them. Specific suppression occurs when select species or groups of beneficial organisms antagonize the pathogen at some stage of its life cycle.

Take-all in wheat or barley becomes less and less of a problem if the crop is grown in consecutive years. Both fungi and bacteria, such as friendly saprophytic *Fusarium* species, reduce pathogen numbers by competing for food supplies, and at the same time specific antagonistic microbes like fluorescent pseudomonads attack the *G. graminis*[192]. The pseudomonads are especially effective when ammonium rather than nitrate fertilizer is used, resulting in a lower rhizosphere pH. This suppression likely occurs mostly in the rhizosphere, but also throughout the soil mass.

In avocado groves of Greensland, Australia, the fungal pathogen *Phytophthora cinnamoni* is controlled apparently by spore-forming bacteria or actinomycetes[193]. These suppressive soils support an impressive array of diverse microbes, unlike other nearby groves that do not have natural *Phytophthora* suppression.

Fungi, Bacteria, and Viruses Attacking Pathogenic Nematodes

Pathogenic nematodes are extremely harmful to a wide variety of crops, causing over a 12% loss to crops worldwide every year[194]. They are attacked by certain types of fungi that produce loops that can lasso the nematodes and digest them (Figure 3-61)[195], or generate a sticky material that glues them to the mycelium. These predacious fungi are especially active when large amounts of organic matter are present in the soil on which they can feed. Bacteria and viruses can also infect and kill or sicken nematodes. Their activity is not confined to the rhizosphere.

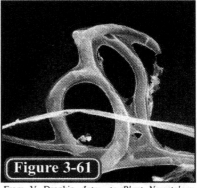

Figure 3-61

From V. Dropkin, *Intro. to Plant Nematology*, ©1989 Wiley Interscience. Reprinted by permission of John Wiley&Sons, Inc.

Nematodes and Mites Attacking Pathogenic Nematodes

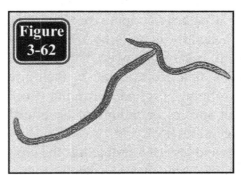

Figure 3-62

Certain species of nematodes are predacious to pathogenic types, and will pierce and consume them (Figure 3-62)[196]. This is much like tigers attacking leopards, but such predation is quite common in soils and is not confined to the rhizosphere. Soil mites, called *mesostigmata* mites, can consume up to 100 or more nematodes each day[197].

Actinomycetes, Fungi, and Bacteria Producing Antibiotics

All of our common antibiotics like Penicillin and Aureomycin have been developed from soil organisms ... especially from actinomycetes. These antibiotics are effective at inhibiting or killing other organisms, sometimes beneficials as well as pathogens. A bacterium called *Pseudomonas fluorescens* produces a phenazine type of antibiotic that inhibits *G. graminis* var. *tritici* — the organism of the take-all — at only 1 u/ml of medium in the laboratory. In the soil environment, wild varieties of *P. fluorescens* are even more effective at controlling take-all than the cultured species. This fact prompted one researcher to state, "Manipulation of the rhizosphere environment to influence production or activity of the antibiotic may increase control of the pathogen[198]".

As the knowledge of beneficial effects of certain rhizosphere organisms to deter pathogens has proliferated, so have efforts to introduce certain biocontrol agents to inhabit the rhizosphere. These agents must be capable of colonizing the expanding root surface after introduction ... i.e., grow along with developing roots. They have been found to protect roots from pathogens in some cases, and also to produce growth-promoting substances and enhance nutrient uptake by roots. One example of a microbial biocontrol agent is *Trichoderma* spp[199]. These bacteria produce large amounts of cellulase, which degrades cellulose on root surfaces and also protects against several pathogens. Inoculating fruit, grape, or rose stock with *Agrobacterium radiobacter* strain K84 protects against *Agrobacterium tumefaciens*, which causes crown gall, a disorganized and uncontrolled division of cells on roots and around the crown. Certain *Pseudomonas* species, applied to roots, can induce resistance to Fusarium wilt[200], and the bacterium *Pasteuria penetrans* can control *Meliodogyne* and other nematodes in certain situations[201].

Helping Maintain Tissue Health to Resist Pathogens

The soil has everything to do with maintaining health of the plant because the soil provides basic building blocks for cell integrity. Nutrients must be provided to the roots within proper limits to insure adequate levels and balance of building blocks at the proper time.

Trophobiosis

One major means by which plants can attract pathogens is to build up an abnormally high level of free amino acids in their tissues. This high level of free amino acids attracts pathogens, which require high levels to build their own proteins; they generally lack the enzyme systems necessary to break down plant proteins for their own use. Thus, they find free amino acids in the vascular tissues and cell contents of unhealthy plants ... those that tend to have higher than normal accumulations of amino acids.

We have already discussed the rudiments of plant metabolism, rhizosphere root exudation of mucigel, and the growth of microbes on the root surfaces. It is important now to understand the essentials of protein synthesis so that we can see how free amino acids are generated. For this, see Figure 3-63[202].

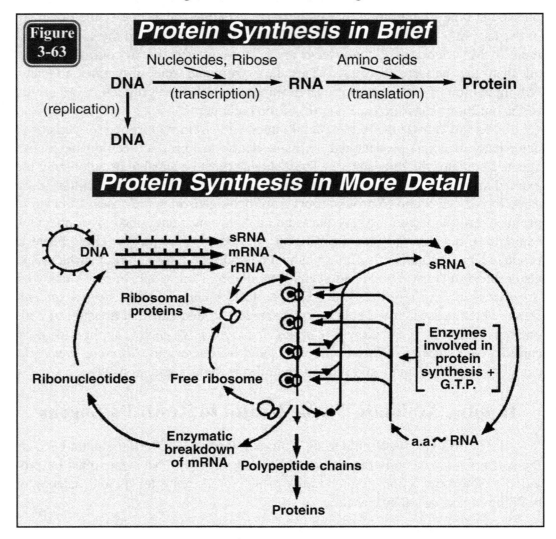

Note that the DNA generates three types of RNA: sRNA (transfer RNA), mRNA (messenger RNA), and rRNA (ribosomal RNA). Each type performs a specific function. The sRNA unites with various amino acids, each special sRNA being specific for a particular amino acid (AA), of which there are over 16. When energized with ATP (adenosine triphosphate), the AA—sRNA then migrates to the long-chain mRMA, where the ribosomes attach and expedite the assembly of polypeptides (chains of amino acids linked together in a definite sequence). Eventually, the mRNA gets old and breaks down, the subunits (ribonuleotides) only to be recycled to form new RNA units.

It is critical to understand that all enzymes are actually proteins having a unique conformation and catalytic quality. Some are called oxido-reductases, which catalyze oxidation-reduction reaction. Others are called transferases (catalyze the transfer of functional groups), hydrolases (catalyze hydrolysis reactions), lyases (catalyze additions to double bonds), isomerases (catalyze isomerization reactions), and ligases (catalyze the formation of bonds with ATL cleavage). In order for enzymes to be active and do their jobs, some of them need no added portion to the molecule, but many require one or both of the following cofacters:

A. Metallic ions, such as Zn^{++}, Mg^{++}, Mn^{++}, Fe^{++} or Fe^{+++}, Cu^{+} or Cu^{++}, K^{+}, Na^{+}, Ca^{++}, Co^{++}, Ni^{++}, or others. These metal ions serve to ...

... act in the primary catalytic center,

... bridge groups to bind the substrate and enzyme together, or

... stabilize the reaction complex to maintain the protein conformation in the active form.

Some enzymes require more than one type of ion to function.

B. Organic molecules, called the *coenzyme*, oftentimes a vitamin

The amino acids that make up the enzyme or protein structures are synthesized through very complex biochemical reaction sequences, which are briefly summarized below in Figure 3-64[203].

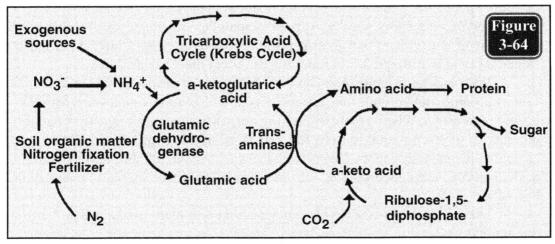

The pathway precisely followed for each of the amino acids differs somewhat, but several are manufactured using the scheme shown above. In all, about 17 major amino acids are produced, while several minor ones are also produced and used in the manufacture of proteins. These 17 are lysine, tryptophan, phenylalanine, threonine, valine, methionine, leucine, isoleucine, glutamine, glutamic acid, proline, alanine, aspartic acid, asparagine, serine, glycine, and cysteine. All have a basic structure as shown below.

$$\begin{array}{c} H \\ | \\ R - C - COO^- \text{, where R is unique for each acid.} \\ | \\ NH_3 \end{array}$$

Why and When Pests Attack Plants

It is important to understand that pathogens of all types have one major characteristic in common: THEIR NUTRITIONAL REQUIREMENTS ARE DIFFERENT THAN NON-PATHOGENIC SPECIES IN THAT THEY REQUIRE RELATIVELY HIGH LEVELS OF FREE AMINO ACIDS IN THEIR HOST TO SURVIVE[204]. They do not possess the enzymes necessary to break down the proteins within the host's tissues, but rather must survive on whatever free amino acids are present in the vascular and cellular fluids. This is because the design and function of these organisms — nematodes, mites, bacteria, fungi, insects, and others — is to eliminate plants which are diseased and sickly ... "nature's clean-up-crew", as it were. These pests are even able to "home in" on the afflicted host through detecting particular wavelengths of infrared and other bands of radiation that the plants emit ... sort of a homing-in signal tuned to the antennae of certain insects, or other sensory mechanisms of nematodes, mites, or even unicellular organisms to notify them that food is present. All living things have a particular frequency, and that frequency is modified to reflect their health status.

With this knowledge, one can now evaluate the means by which a plant's tissues become glutted with free amino acids, the food supply necessary for pathogenic organisms. When a plant is growing normally as programmed by its genes, the supply of N-compounds from the soil is adjusted to the needs of the proteosynthetic machinery in the leaves; no excess will occur, but the supply of amino acids is immediately incorporated into proteins by an active synthetic pathway in leaf cells.

Excessive amino acid levels can accumulate in plant tissues when the following occur[205]:

1. Too much nitrogen is available in the root zone. This occurs by applying moderate to heavy fertilizer nitrogen or manure applications. Nitrate and NH_4^+ are absorbed and incorporated into amino acids, which become overabundant in relation to the cell's ability to utilize them (Figure 3-65)[206].

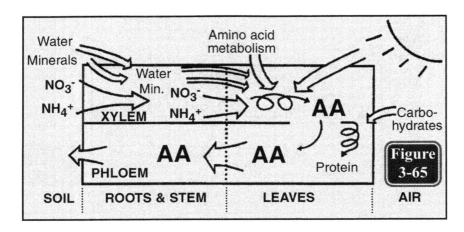

Figure 3-65

2. Proteosynthesis is slowed, due to...

a. Mineral shortages, both micronutrient enzyme activators and macronutrients; these partially active proteosynthesis enzymes are then unable to utilize a normal flow of amino acids, creating a backlog that spills over into plant tissues (Figure 3-66)[207].

b. Pesticide applications. Pesticides act as enzyme poisons for protein synthesis and slow the incorporation of amino acids into proteins, creating a backlog that spills into other plant tissues (Figure 3-66)[208].

c. Dry conditions, which reduce the uptake of minerals, increase stress, and reduce enzyme activity, which interfere with protein synthesis; the result is an amino acid backlog in tissues (Figure 3-66)[209].

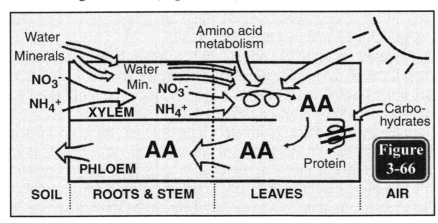

Figure 3-66

3. Recycling of mobile nutrients, including amino acids, to younger growing tissues. Two cases are common:

a. Older plant tissues senesce, and proteins break down to their constituent amino acids and are transported out of the older tissue to the younger tissue (Figure 3-67)[210].

b. When soil nitrogen levels are inadequate, proteins in older leaves break

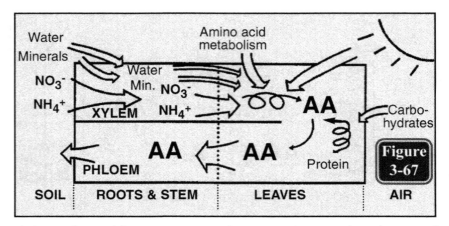

Figure 3-67

down and the amino acids are transported to younger, growing tissues. Only the older, senescing tissues where proteolysis is active may contain high free amino acid levels (Figure 3-67).

In summary form, the means by which high free amino acid levels are achieved are illustrated below.

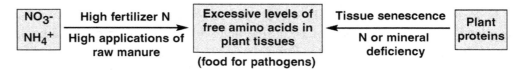

In all of the above cases, high free amino acid levels result in plant cells and vascular tissues of at least parts of the plant. This leaves the plant susceptible to pathogen attack: nematodes, viruses, fungi, and bacteria in roots, and mites, bacteria, fungi, viruses, and insects in leaves.

By building up high free amino acids in plant tissues through excessive nitrogen applications, inadequate minerals (enzyme catalysts), and pesticide applications (enzyme poisons) — all of which are under the control of farmers — plants are made susceptible to pathogen attack.

Nitrogen compounds and mineral elements enter the roots through rhizosphere microbial mediation, as discussed earlier, so one can readily see that rhizosphere nutrient conversions — as much as friendly pathogen antagonists — are responsible for plant health and its maintenance. Pesticides are introduced by farmers, so their role in plant disease control cannot be minimized, but their role is not closely tied to the rhizosphere mechanisms already discussed.

Root Community Relationships

Mycorrhizae

Perhaps no other association of organisms within the plant-soil system is more profoundly illustrative of the Creator's character than the beautiful mutualism expressed in mycorrhizal connections among plant roots. Mycorrhizae not only greatly benefit the individual plants with which they are associated, but they can — and usually do — reach out to attach to the roots of other plants in the vicinity, feeding nutrients from the stronger plants to the weaker ones, boosting the overall health of the entire plant community in a direct and substantial way. This is the epitome of love and selflessness just as directed within human associations: "Do unto others as you would have them do unto you, which is the whole meaning of the law and the prophets"[211]; "you shall love your neighbor as yourself"[212]. The selfless works a person does for others — not letting the right hand know what the left hand is doing[213] — are the most valuable one can do, especially those done in secret. What incredible similarity this concept has with the highly beneficial mycorrhizal activities occurring within the dark and hidden recesses of the soil[214]. Matthew 25:35 directly states that,

> "... I was hungry, and you gave me food; I was thirsty , and you gave me drink ..."[215].

The mycorrhizal fungi, through their power to connect plants through their vast mycelial network, fulfill the very functions of the righteous sheep positioned at the right hand of the Son of Man at His second coming. This profound message found within the operation of the natural world beneath our feet is no accident. It was placed there by the Creator and is a direct reflection of His character.

Radioactive isotope studies by Simard and others[216] in Canada have shown that different species of plants can be compatible with the same species of mycorrhizal fungi. Their studies showed that the transfer of carbon compounds was in both directions, between *Betula papyrifera* and *Pseudotsuga menziesii*, but the transfer was mostly towards *P. menziesii*, especially when *P. menziesii* was shaded ... meaning it was not fixing as much carbon as before and needed help from somewhere. The transfer of energy to assist *P. menziesii* in its trouble was contributed by *B. papyrifera* and transmitted by the mycorrhizae.

Notice the marked transfer of radioactive phosphorus ($^{32}PO_4$) from the source area in the lower right of the growth container in Figure 3-68. The radioactivity shows up as dark patches along the root and mycelial network, moving up from the source and along the mycelia and to various resource-rich patches that interconnect plants. Even the leaves of the pine seedlings show up in the auto radiograph.[217] Studies like this show that in ectomycorrhizal plants such as pines, it is the mycorrhizal mycelia rather than the roots that are primarily responsible for the absorption and transport of nutrients amongst interconnected plants.

When radioactive carbon ($^{14}CO_2$) is fed directly to a "donor plant" in an array of several plants, it has been discovered, through autoradiography, that the carbon becomes distributed throughout the root systems and mycelia of all of the plants that are interconnected by the mycorrhizae. In the auto radiograph shown in Figure 3-69, it was subsequently discovered that the patches of radioactive activity (see the two arrows) are mycelia that are involved in the mobilization of nitrogen.[218]

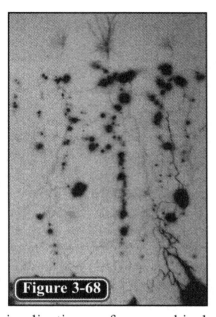

Figure 3-68

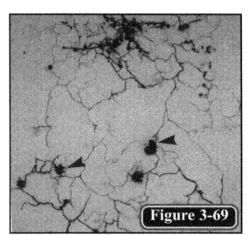

Figure 3-69

The implications of mycorrhizal links and interplant carbon and nutrient transfer for ecosystem structure and dynamics are numerous. Some researchers argue that the nutrient pathway — hyphal versus soil — is irrelevant from practical and economic viewpoints, but at the same time say that understanding these mycorrhizal pathways is important for plant management.[219] Some of the important implications of mycorrhizal links amongst plants are given below.[220]

1. Mycorrhizal associations assist seedling establishment near mature plants by allowing the seedlings to become infected more rapidly, or to tap into an established common mycorrhizal network supported by other plants. Note the picture in Figure 3-70, where the emerging radi-

Figure 3-70

cle of a newly germinating *Pinus* seedlings is becoming incorporated into the mycelial network growing from, and supported by, the established central plant.[221] These plants incorporate into what have been called "guilds", with all of the associated plants interconnected by a common mycorrhizal fungus.

2. Mycorrhizal associations assist the recovery of species following disturbance of the soil or of the plant stand.[222]

3. Mycorrhizal associations reduce competitive dominance of one species, and promote the diversity of species within an ecosystem by allowing carbon and nutrients to flow directly through the mycelial network from well-suited to deficient plants nearby.[223] Since a diversity of plants at a location is normally more beneficial for plant health and resource utilization than is a monoculture, the mycorrhizal network thus aids materially in establishing and maintaining an improved ecosphere. The strong literally assist the weak and poor individuals, a highly spiritual principle.[224]

4. Mycorrhizal associations reduce nutrient losses from ecosystems by keeping more nutrients in the biomass, thereby increasing overall productivity. This loss reduction is especially evident when nutrients are cycled from dying to living plants.[225]

5. Mycorrhizal associations increase the productivity, stability, and sustainability of ecosystems, an inferred effect that has never been treated experimentally.[226]

Due to the many positive impacts of mycorrhizal hyphae connections among plants to create a more stable, efficient, healthy, and diverse total ecosystem, the implications of activities such as tillage and pesticide applications (herbicides, insecticides, fungicides, etc.) on overall health of the ecosystem are great. Massive fracturing and mixing of the soil by machinery from the perspective of hyphal interplant connections is highly deleterious to plant and soil health. These community sensitivities are a tribute to the need to treat soils with care, and protect and respect the established habits and structure of interplant organisms. Even as one should not destroy the contacts between citizens of a town without expecting serious upheavals in its function, so one should not disrupt the contacts among plant root systems by rupturing the multitudes of interconnecting fungal hyphae.

Other Forces and Organisms

The plant community communicates in other ways besides mycorrhizal intercommunications ... both above and below-ground. For years it has been understood that leaves of trees which are damaged by tearing send some sort of airborne cue to surrounding trees, telling them to prepare for an insect or pathogen attack. In one experiment, poplar leaves were damaged, which led to an increased synthesis of protective phenolic compounds and certain tannins — which are harmful to predacious insects — and within 36 hours undamaged nearby trees exhibited similar responses.[227]

The means of communication between plants is much more sophisticated than previously thought, according to recent studies. Individual cells act as both receivers and transmitters, and cells, tissues, and entire organisms each carry specific frequencies of either electromagnetic energy or some other type of energy. For instance, the lowly *E. coli* bacteria has been found to possess at least four major surface chemoreceptors which sense specific compounds in the environment. These chemoreceptors then let the cell know whether to move towards or away from a chemical signal ... i.e., towards a food source or away from a damaging chemical.[228] These sensors work together to sense a particular compound and evaluate its worth to the life of the cells.

The sensory potentials and movements of plants are much more extensive than most people think, even to the point that plants might be perceived as having minds of their own. One gifted Austrian biologist named Raoul France' even forwarded the idea that plants move their bodies as freely, easily, and gracefully as many animals and humans, albeit at a much slower pace.[229] He said that plants are capable of intent: they can stretch toward or seek out what they want in strange ways. As France' pointed out, far from existing inertly, the inhabitants of the pasture appear to be able to perceive and react to whatever is happening in their environment at a level of sophistication far surpassing that of humans.[230]

Professor Harold Burr in the 1930's and 1940's, at the Yale Medical School, carried out some amazing experiments into the energy fields surrounding plants, trees, humans, and even individual cells.[231] These fields must surely relate to the ability of plants and their roots to communicate with other plants in their surroundings, and even to microorganisms that populate the rhizosphere.

A researcher from I.B.M. worked with plants for several years, discovering that their "emotions" or energy fields could be recorded on sensitive equipment that measures electrical output ... such as devices used to detect lies in people. He found that feelings of joy, contentment, hurt, or pain all produced specific response patterns on his equipment. He concluded that,

> "It is fact: man can and does communicate with plant life. Plants are living objects, sensitive, rooted in space. They may be blind, deaf, and dumb in the human sense, but there is no doubt in my mind that they are extremely sensitive instruments for measuring man's emotions. They radiate energy forces that are beneficial to man. One can feel those forces! They feed into one's own force

field, which in turn feeds back energy to the plant"[232].

The ability of plants to communicate with other plants, and with people, animals, and with virtually any living cells or tissues has been quite thoroughly elucidated through the work of Cleve Backster. A polygraph examiner and teacher of lie detection methods to students from around the world, he pioneered the measuring of plant reactions to various treatments — such as leaf damage and uprooting — and even to thoughts towards these plants. He discovered that plants connected to polygraph leads would display alarm responses even to thoughts of *intended* harm, but only when the perpetrator was truly intent on committing such harm. The plant could determine whether the thought was genuine or not — not a pretended intent — so in a sense the plant could function as a witness to the truth of a person's very thoughts[233]. Backster had a friend enter a room having two plants, one of which the person destroyed. Later, with a polygraph connected to the survivor, and with several others in the room, Backster was able to easily determine which of the friends had killed the first plant by the frantic output its surviving mate put out on the polygraph in the proximity of the culprit[234]. Plants can indeed communicate in unseen ways with the world around them, and in many ways sense and reflect the emotions of people.

Philip Callahan, a former United States Department of Agriculture entomologist, has shown that insects communicate with one another and with plants via radio frequencies and pheromones[235]. The antennae of insects are shaped to detect specific frequencies essential for food-getting or reproduction ... detecting faint chemical emissions from food sources, or the location of a mate that may be several miles away. These antennae are also linked to the ability to detect sick and ailing plants, which put out a very different electromagnetic spectrum than healthy plants. Thus, the ailing plants attract the insects that are programmed to feed on them. The infrared band of the spectrum is an especially important band used by

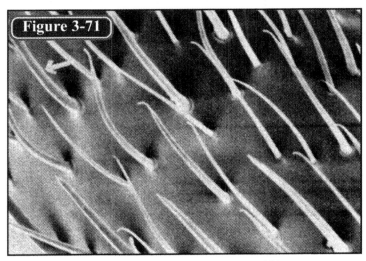

Figure 3-71

insects. Note the tapered sensilla on the antenna of the fire ant in Figure 3-71. The ants vibrate their antenna at from 12 to 20 cycles per second while following a trail to modulate "maser-like" frequencies from the trail scent[236].

There are without a doubt very intricate and subtle interactions amongst the various cells, tissues, and organisms that live within the soil. As already discussed, people and insects interact with plants; surely virtually all living cells similarly interact in some way. Not only are the mycorrhizae involved in these intercommunications, but the whole host of soil microbes that emit and receive messages from unseen energy wavelengths ... from the electromagnetic spectrum or from other energy sources not yet well comprehended. How, after all, are "attitudes" transmitted from plants to the various microbe species in the root zone? It is thought by some rhizosphere researchers that the plant has an intelligence to selectively feed, and thus encourage the growth of, specific microorganisms in the root zone to benefit the plant's immediate needs[237]. An example might be to encourage a certain type of actinomycete that generates an antibiotic against a virulent *Fusarium* species that is attacking the roots. The signal to encourage the actinomycetes might be an array of amino acids, carbohydrates, and vitamins that the plant's root would secrete into its "microbial garden", thus stimulating the actinomycete.

The symbiotic array of roots and organisms within the soil is indeed a complicated one, but is surely based upon the same principles of mutualism that the Creator intended to operate within the human spectrum and throughout all of the creation. Love and consideration for one's fellow man, and supplying for the needs of others when they are needed, is a key to the proper functioning of any living system. The soil and the root systems of plants are no exceptions.

By understanding these amazingly complex and interactive mutualistic associations among plants, it is possible to envision why in many cases monoculture plantings are so susceptible to disease and insect attacks ... why large plantations of bananas, pecans, sugar cane, or corn must so often rely on pesticides to control outbreaks of plant-pathogenic nematodes, mites, insects, fungi, or bacteria. If the food crop being grown is not receiving the required minerals, growth regulators, vitamins, antibiotics, and other important growth factors through the mycorrhizae and other subterranean interconnections as supplied by other plants in the vicinity, then the crop is vulnerable to pathogen attack. As the Creator designed things, protection comes from the nearby support of those having different and diverse abilities, as it were. The stronger truly support the weak, and there is strength in diversity.

Chapter IV
How to Manage Soils for Lasting Productivity and Compatibility with Nature's Laws

We have just reviewed a great deal of information about what soils are and how they function. It is clear that soils are living bodies that cover the earth as a thin skin ... the six inches that separate life from death. It is clear that our lives depend on the soil's prosperity, since mankind survives on the produce of the land, be it plant or animal life.

These few obvious facts make it clear that, in order for people to inhabit the earth in prosperity for the foreseeable future, we must carefully manage our soils to support such sustained productivity. The manner in which we do that must take into account what soils are and how they operate. Thus, we must seek to optimize and improve upon *what is*, not engage in conflict with the created systems around us.

Basic Principles

We are thrust into the task of discovering and supporting the natural laws by which soils operate. They can be summarized briefly by four points put forth by Japanese farmer-philosopher Masanobu Fukuoka in *The One Straw Revolution*[1]. Note that these points are the ideal, and do not restrict a farmer from following practices that are biologically sound and bring the soil into an optimum state of fertility over a shorter time span.

1. **No cultivation**. The earth cultivates itself naturally by means of the penetration of plant roots and the activity of microorganisms, small animals, and earthworms.
2. **No chemical fertilizers or prepared compost**. If left to itself the soil maintains its fertility naturally, in accordance with the orderly cycle of plant and animal life. Only plant residues (straw, green manure crops, and so forth), animal wastes, and other nonsynthetic materials should be added to soils, and then in the proper amount and at the proper time as in a natural system.
3. **No weeding by tillage or herbicides**. Weeds should be controlled but

not eliminated, such as by straw mulching, temporary flooding of rice fields, and timely management practices.

4. **No dependence on chemicals**. Nature left alone is in perfect balance. Harmful insects and plant diseases are always present, but do not occur in nature to an extent that requires the use of poisonous chemicals.

One might object that these four points are not workable within the modern paradigms of industrial agriculture. This is indeed true, but be aware that the modern paradigms of industrial agriculture are doomed to fail ... and sooner than we might think. Industry's tenets of profit maximization run roughshod over nature's tenets which adjust yields of native, adapted crops to local soil fertility, without help from industrial inputs such as anhydrous ammonia, superphosphate, triazines, or organophosphates. Industry's historical record proves its bent for short-term gain and exploitation of resources rather than receiving the natural productivity of the soil while returning at least as much as was removed. The bent of industrial ethics has been to maximize short-term gain while neglecting the long-term care of resources. Such views are incompatible with the optimum health of living systems. They may force soils to produce crops which appear to be abundant, but are deficient in the proper levels and balance of nutrients ... while harboring traces of cancer-causing pesticides and allowing erosion to insidiously wash away fertility year by year, lowering prospects of future yields. Optimum profits are sought after on the land, not optimum health of the consumer ... whereas optimum health ought to be the number one objective of any exploit of mankind.

Understanding this basic defect in society's approach towards agriculture, one can then justify a degree of ultruism in treating the land. Indeed, the law of love applies to — or ought to apply to — our management of the soil. We ought to treat it as we ourselves would like to be treated ... with respect, honesty, kindness, humility, and generosity. We ought to treat it as the medium of life-engendering growth for us and our families and nation, not just for today but for millennia to come if we really care for our children and our children's children. Besides, we ought to consider the *health* of our families that results from the food our soils produce. While the strong, immutable connection between soil health, plant and animal health, and human health is beyond the scope of this work, it should be appreciated that the soil health-human and animal health link is a powerful one. As we do to the soil and the crops that grow on it, so it is done onto us ... "in good measure, pressed down, and shaken together, and running over ... 2".

We are left with few options when attempting to understand how to properly manage the soils upon which we survive than to carefully observe the natural world our Creator put into play. What do we see when standing amidst a sea of tall prairie grasses, or beneath the canopy of a towering forest nestled away from

126

the undue influence of man? We see few or no nutrient deficiency symptoms. We see healthy, deep green leaves which have few worm holes or galls. Insect pests pose no serious problem, nor do root nematodes or fungal diseases. This does not mean that leaf-eating grasshoppers, sap-sucking, virus-transmitting aphids, or root destroying *Meloidogyne* nematodes are not present. They are, but their populations are held in check by friendly predators of these pests: ladybugs, golden-eyed lacewings, nematode trapping fungi, and milky spore fungi. Pests are also discouraged by healthy plants that defend themselves against the invasions of pathogenic fungi, nematodes, bacteria, and viruses, and harmful insects. Not only are free amino acid levels low in tissues because the enzyme machinery rapidly incorporates the amino acids into proteins — thus discouraging pest feeding — but tissues can build up toxic substances, such as alkaloids, to insect pests and discourage their feeding. The plants can even somehow communicate an insect or pathogen invasion to neighboring plants so they can prepare a defense.

While the earth cultivates itself naturally through the actions of roots and small animals and insects, maintains its own fertility by cycling residues, fixing nitrogen, and making more nutrients available, and maintains a balance between harmful and beneficial insects and plant diseases, there is still plenty the farmer can do to enhance the systems already in place. He usually is forced to do so through sheer economic pressures to survive in today's competitive economic system. Yet, it is possible to survive quite well while closely emulating points 1, 2, and 4 of Fukuoka. Point 3 is a bit more problematic.

Long-Term Field Studies

To prove this point, two reports from the prestigious journal *Nature* show how an agriculture oriented towards the laws of nature really works in today's world[3]. A 15-year study by the Rodale Institute (USA) examined three different systems:
1. **Conventional**. Typical commercial fertilizer and pesticide inputs in a corn/soybean rotation
2. **Organic with livestock**. Manure used as the sole fertilizer source for corn, from cattle fed on the corn, and no pesticides
3. **Organic without livestock**. Legumes for nitrogen production with corn, and no commercial fertilizers or pesticides

Over a 10-year period the three systems showed little difference in yield (less than 1%), and were about equal in profitability. However, the organic with livestock system was significantly superior in soil fertility to the conventional system, and a bit above the organic system without livestock. The conventional system reduced soil fertility, and had allowed 60% more nitrate to leach into the groundwater than the other systems.

The second study is a review of 150 years of work at the Rothamsted Broadbalk Experiment Station in England[4]. Wheat yields on manured plots averaged 3.45 tonnes/hectare, while commercial NPK plots yielded an average of 3.40 tonnes/hectare.

Moreover, total soil nitrogen and soil organic matter increased by 120% over the 150 years for the manured plots, but by only 20% for the commercial NPK plots. Clearly, manure plays a highly important role in the improvement of these English soils, while maintaining yields at high levels.

These long-term studies, especially the English one, show that mixed farming with livestock, with manure returned to the field, greatly benefits soil organic matter and fertility. Conventional NPK fertilizers did not improve the yield and profit picture versus "organic" methods for the Rodale Institute study in the United States. These are extremely important studies supporting the notion that emulating nature's laws on the land does indeed make financial sense while benefiting soil properties.

Reduction of Tillage

The studies mentioned in *Nature* did not reduce tillage, however, which contradicts Fukuoka's first point. Tillage using the moldboard plow was soundly denounced by Edward Faulkner in *Plowman's Folly*[5], wherein he stated,

"... the moldboard plow ... is the least satisfactory implement for the preparation of land for the production of crops .... The entire body of 'reasoning' about the management of the soil has been based upon the axiomatic assumption of the correctness of plowing. But plowing is not correct .... Had we not originally gone contrary to the laws of nature by plowing the land, we would have avoided the problems as well as the expensive and time-consuming efforts to solve them."

Faulkner approved of mixing residues into the surface with a disk harrow, not inverting it with a moldboard plow. This was not true minimum tillage, much less zero tillage as oftentimes practiced today, but it was a step in the right direction. Less soil erosion resulted because residues remained near or on the surface, and a plow pan was avoided. Implements were lighter in the early 1940's as well, and disk harrows were smaller and did not penetrate and compact the soil as greatly as do many modern tractors and disks.

The problem with artificial tillage is many-faceted. Soil structural units are crushed and broken, exposing the protected internal organic matter to rapid breakdown. Thus, organic matter levels drop quickly as, with a flush of oxygen to power microbial activity, the organic materials break down to CO_2 and their constituent minerals and nitrogen. Crops utilize these nutrients and do well for several years — perhaps decades — until the easily decomposed organic matter dis-appears. Then artificial fertilizers are needed to boost production, but these tend to further accelerate microbial growth and break down additional organic matter. More insidiously, the exposed soil is subject to direct raindrop impact and dispersion (Figure 4-1), allowing the soil to erode much more easily than on surfaces protected by impact-absorbing trash or living plants. The lighter, more fertile organic materials wash away first, leaving the less fertile, denser mineral matter behind, as shown in Figure 4-2[7].

Figure 4-1

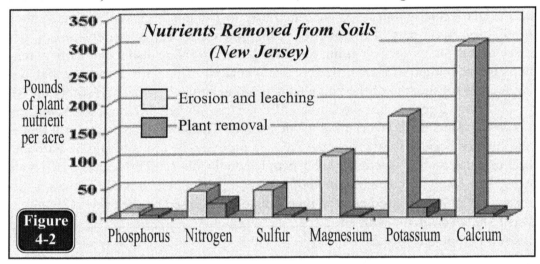

Figure 4-2

Nutrients Removed from Soils (New Jersey)

Pounds of plant nutrient per acre

☐ Erosion and leaching
■ Plant removal

Phosphorus   Nitrogen   Sulfur   Magnesium   Potassium   Calcium

Wind erosion is no less serious than water erosion in vulnerable areas, such as the Great Plains of North America, In one sampling of windblown dust originating from Texas and Oklahoma and sampled in Iowa, 500 miles away, the enrichment factor of the eroded soil was great[8]. Note Figure 4-3.

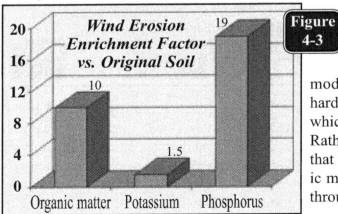

Figure 4-3

Wind Erosion Enrichment Factor vs. Original Soil

10   1.5   19

Organic matter   Potassium   Phosphorus

Tillage which we practice in modern Western agriculture is hardly a mutualistic process by which good is done to the soil. Rather, it is a parasitic process that oxidizes vital stores of organic matter and accelerates soil loss through erosion.

Biblical Principles

The Biblical injunction to "till" the soil comes from the Hebrew word *abad*, meaning "to serve, labor, or work"[9], while the word "keep" comes from the Hebrew *shamar*, meaning "to keep, observe, or take heed"[10]. Notice that the command for Adam to "tend [dress] and keep" the Garden of Eden[11] did not involve laborious inversion of the soil, or even breaking up of the soil surface. Rather, it involved *service* to the soil and the plants that grew there, which was not sweat-provoking, difficult work; that curse was given later: "In the sweat of your face shall you eat bread ..."[12].

In the Garden of Eden the work of Adam did not involve tillage as we know it, but rather serving the growth and health of perennial plant species. God ordered Adam to "... freely eat of every tree of the garden ..."[13]. Trees are woody perennials. This is not to say that the order to use green plants bearing seed as food[14] was not a part of the plan for man's provision before sin entered in, but these herbal plants such as grain, root, and leaf crops were also most likely perennials in the Garden of Eden ... before death entered in: "... for in the day that you eat thereof [of the tree of the knowledge of good and evil] you shall surely die"[15].

Once forced out of the Garden of Eden things were different, for Adam and his descendants had to work hard for their food. The soil would bring forth thorns and thistles besides food crops, the work to produce the crops would be hard and provoke sweat, and eventually the people would die[16]. This change of locale caused an essential change in methods of food production as well. However, the means of "plowing" used early in man's history, and used by the ancient Hebrews, was to *make a furrow* with an instrument pulled by an ox (*plowman* = Hebrew *charash*, "to make an indentation")[17]; see Figure 4-4[18]. Seeds were then dropped into the furrow and covered by a second person, or, as shown in Figure 4-4, a drill tube was used. The entire soil surface was not fractured as in plowing today.

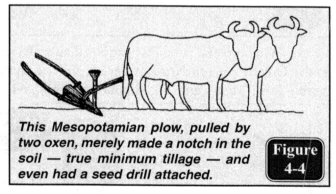

This Mesopotamian plow, pulled by two oxen, merely made a notch in the soil — true minimum tillage — and even had a seed drill attached.

Figure 4-4

It is apparent that the ancient Hebrews practiced nearly a zero form of tillage. That system worked for thousands of years, and works today as well through several systems of planting. However, herbicides are usually used for weed control. By not inverting the plow layer the residues are retained on the surface, providing insulation against temperature extremes, reducing weed emergence, adding nutrients and humus during breakdown, and providing a barrier to moisture loss. In the process of organic matter degradation, a strong soil structure is fabricated as micro and macroorganisms proliferate on the surface few inches of soil during the breakdown and incorporation process. This is the way nature operates. It is the way farmers should emulate in their own management systems.

It is not essential to apply commercial NPK fertilizers to produce abundant crops today. The studies cited earlier in *Nature* prove that. Organic farmers across the United States are likewise getting excellent crop yields without commercial fertilizers. Instead they are utilizing manure, compost, and residues from green manure crops and the previous crop for basic nutrient sources. Crop rotations with legumes such as alfalfa help maintain structure and nitrogen levels, though tillage is required to turn under the legume. Natural ground rock minerals such as lime, granite, rock phosphate, potassium salts, and sea salts supply major and minor nutrients. Aggressive growth resulting from such a system of management implies healthy plants that naturally repel insect and microorganism pests.

Weed control for annual crops is next to impossible under non-chemical schemes without cultivation of some sort. Even organic gardeners resort to hoeing out weeds if a mulch has not smothered them. Thus, few organic farmers growing annual crops can utilize no-tillage methods of crop production because herbicides are not allowed. This fact points to the need to develop perennial varieties of corn, wheat, barley, soybeans, and other annuals that can be established and can then fend for themselves amongst weeds in successive years, with only minor non-chemical weed control, such as hand pulling of weeds. The weed problem also points out that perennial tree and fresh crops are desirable — apples, peaches, cherries, plums, citrus, mangoes, pineapples, blackberries, bananas, walnuts, Brazil nuts, and a host of others in every part of the world — to obviate serious weed problems so often encountered with annual crops.

Unfortunately, seed companies zealous for profits encourage the use of annuals so they can sell the farmer seed year after year. These companies have gone even a step further, by genetically altering the crops to tolerate normally lethal herbicides, and thus sell the "patented" seeds along with the herbicide to kill weeds. Some companies have even gone as far as to produce seeds that will not germinate the following year, preventing the farmer from planting his own seeds and requiring him to return to the seed producer. The greedy schemes of multinational corporate agriculture have few limits. Yet, the result of this attempt to box farmers into purchasing genetically altered seed only from certain companies carries the threat of famine should for some reason the companies be unable to supply the seed, and the farmer must replant what seed he has ... which will not germinate. Even more ominously, the health effects of consuming genetically altered grain have not been explored. Humanity sits on a time bomb that is set to explode upon a people whose greed has led them to value money more than life and happiness.

Major Practices to Follow

The direction that farmers ought to take, as a result of the above considerations, is quite simple: follow those practices that encourage the establishment of a well-developed rhizosphere that is capable of maximum nutrient release and crop growth, and builds a stable, strong soil structure within a soil whose organic matter content is maximum for the locale. This implies vigorous microbial and earthworm populations that conspire to maximize the mutualistic interactions of soil bacteria, fungi (especially mycorrhiza), cyanobacteria, algae, protozoa, actinomycetes, and other forms of soil life. To achieve this yield-maximizing soil environment, the following practices are recommended.

1. **Minimum or zero tillage**. Allow crop residues, manures, fertilizers, and other nutrients to collect on the soil surface and work into the soil through the action of soil organisms. This may require moving towards perennial and tree crops if they will sustain one's livelihood.

2. **Balanced soil fertility**. Test the soil to determine deficiencies or imbalances, and then add nutrients from organic or inorganic sources to achieve a fertility goal. A good system of fertility management ought to maximize rhizosphere microbial activity so plants will be fed the optimum levels of nutrients at the proper time. An excellent approach to such management is using the soil mineral balancing concept pioneered in America by William Albrecht[19]. In this system, the major soil cations (calcium, magnesium, potassium, sodium, and hydrogen) are balanced in terms of percent base saturation as shown in Figure 4-5. Other nutrients (phosphorus, nitrogen, sulfur, iron, copper, etc.) are calculated in terms of sufficiency levels to achieve the desired yield. This does not mean that a soil well outside of these limits cannot be productive if care has been given to insure a vigorous soil biology and high organic matter level. Soil nutrient deficits can be cor-

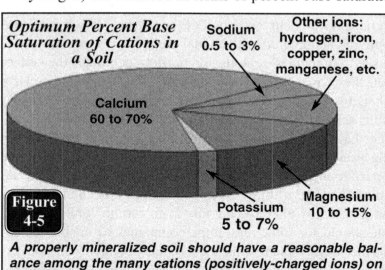

Optimum Percent Base Saturation of Cations in a Soil

Sodium 0.5 to 3%

Other ions: hydrogen, iron, copper, zinc, manganese, etc.

Calcium 60 to 70%

Potassium 5 to 7%

Magnesium 10 to 15%

Figure 4-5

A properly mineralized soil should have a reasonable balance among the many cations (positively-charged ions) on the exchange complex. With magnesium in the 12 to 15% range, soil properties will tend to be optimized provided other elements, especially calcium, are within their appropriate ranges.

rected by natural ground rock minerals such as limestone, rock phosphate, and granite. Truly complete fertilizers are organic in nature, containing not just nitrogen, phosphorus, and potassium, but significant levels of other macro and micronutrients as well as high levels of carbon compounds to replenish organic matter stores. Applications of 3 to 10 tons per acre of manure, or 2 to 5 tons per acre of compost, provide excellent supplementation for high yields.

3. **Natural weed control**. Perhaps "natural" is not the correct word to use here, but the word implies not using toxic chemicals to control weeds. Rather, machine cultivation, hand picking, or flame cultivation can be used. For large acreages of annual row crops like corn and soybeans, flame cultivation avoids tillage, whereas conventional cultivation disturbs the soil. Most farmers cannot pick weeds by hand except in garden plots, while hoeing tills the soil essentially as does cultivation. Hoeing is used extensively in developing nations such as Ecuador, Indonesia, and Sub-Sahara Africa. Cultivation of any sort disturbs rhizosphere establishment, thus inhibiting nutrient supply to the plant, especially when soluble NPK fertilizer sources are used. Yet, there is oftentimes no alternative in this age to cultivation for weed control.

    By growing perennial crops such as hay or pasture, mowing and grazing largely control weed populations without cultivation. Orchards and vineyards can utilize cover crops or permanent grasses between rows. Forestry settings require essentially no weed control at all, nor do natural prairie and grassland settings once they are established.

4. **Natural pest control**. Healthy plants, as we have seen, resist pathogen attack by having low levels of free amino acids and the ability to fight back strongly through inborn immune responses. Healthy plants grow on healthy soils, those replete with abundant and active organic matter which contains a wide array of small and large beneficial organisms to fight pathogenic nematodes, fungi, and bacteria that may threaten roots. Insect and microbial pests flourish when high nitrogen levels are present from heavy synthetic nitrogen applications. Plants are encouraged to build thin cells walls which are vulnerable to pests, and to increase free amino acids in tissues which also increases their vulnerability. Other nutrient imbalances likely produce other metabolic disturbances that upset plant metabolism and encourage pests.

    An array of different plant species within the same proximity has been shown to reduce or eliminate root and leaf pest problems. The reasons for this have already been discussed for *Gros Michel* bananas in natural forest settings, but the same principle applies to all crops. A mixed legume-grass crop has fewer insect pest problems than does an all-legume or all-grass crop, and mixed cultivations — such as corn amongst fruit trees — tend to yield more than straight monocultures[20]. The reasons for this benefit of a mixed plant setting may have to do with nutrient sharing in the root zone (a community of plants can graft roots one to another and/or interconnect their mycorrhizae and share nutrients), superior nutrient uptake and thus improved pest immunity, a better light spectrum, improved water relations, or simply a lack of soil disturbance so that a well-established rhizosphere will provide nutrients and water at rates the plants can use them ... and not glut or deprive plants of their needs. No pesticides mean no interference with metabolic cycles, few excesses of free amino acids, and little destruction of beneficial soil microbiota.

5. **Land use according to capability**. Soils have inherent capabilities for productivity. While these limitations can be improved upon by mineral and organic matter supplementation, the limitation of slope when tillage is used cannot be avoided. Soils will erode on hillsides, even those that have only 1 to 2% slope, if raindrop action can disperse exposed soil.

   Various land capability classes have been used for years in soil survey work. Eight classes are commonly used, which are depicted in Figure 4-6[21].

   Class I: requires good management practices only; suitable for cultivation
   Class II: moderate conservation practices necessary; suitable for cultivation
   Class III: intensive conservation practices necessary; suitable for cultivation
   Class IV: perennial vegetation recommended, with infrequent cultivation
   Class V: no restrictions in use for pasture, hay, or woodland
   Class VI: moderate restrictions in use for pasture, hay, or woodland
   Class VII: severe restrictions in use for pasture, hay, or woodland (highly sloping)
   Class VIII: best suited for wildlife and recreation (rocky hillsides, very rough

Figure 4-6

Class VIII Land

Class VII Land

Class VI Land

Class IV Land

Class II Land

Class V Land

Class I Land

Class III Land

terrain)

6. **Soil conservation practices used whenever needed**. Row crops with exposed soil should be grown only on Classes I, II, and III, and on Classes II and III only if appropriate soil conservation practices are used. These include one or more of the following as depicted in Figures 4-7 through 4-10[22].

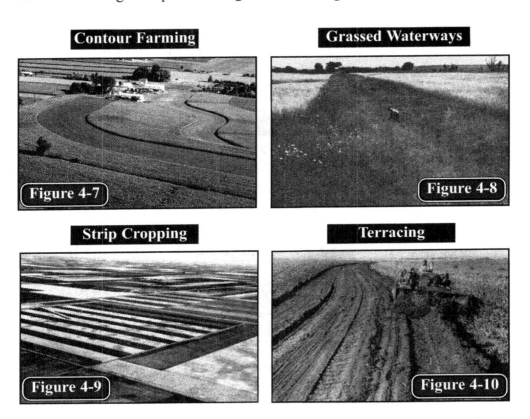

Contour Farming

Figure 4-7

Grassed Waterways

Figure 4-8

Strip Cropping

Figure 4-9

Terracing

Figure 4-10

Shelterbelts are most important for areas of wind erosion. So are tree plantings for rough areas and formerly gullied ravines that need stabilization. The encouragement of wildlife by restoring native adapted vegetation is always an appropriate practice if the soil is marginal for crop use.

7. **Mixed crops and livestock**. The natural world teaches us that animals coexist with the plant world. When the white settlers arrived in America there were perhaps 50 to 100 million buffalo roaming the country, mainly in the area of the Louisiana Purchase[23]. Countless more millions of pronghorn antelope, elk, mule deer, whitetail deer, moose, mountain goats and sheep, and other animals roamed the continent. The present-day Corn Belt, Great Plains, South, and East all had their native livestock species which grazed the land, deposited manure, and completed the cycle of soil to grass, back to soil again.

When the European settlers arrived in America many of them took up homesteads across the middle and western portions of the nation. Though owning land without cost, they were required to till and plant a certain portion of the land each year, and build fences ... a small price to pay for one's own kingdom in a vibrant, fertile land. Nearly all of these settlers owned and bred livestock: horses, cattle, sheep, goats, hogs, or mules. At this juncture of history many nutrients were recycled back to the soil as manure, although tillage with the moldboard plow was nearly universal, especially on the grasslands. Grains, however, were exported off the land to the cities in a one-may nutrient flow.

For many astute managers who appreciated the worth of their land and intended to hand it down to their sons, recycling of nutrients through manure and crop residues was critical (Figure 4-11)[24]. Their land produced abundant-

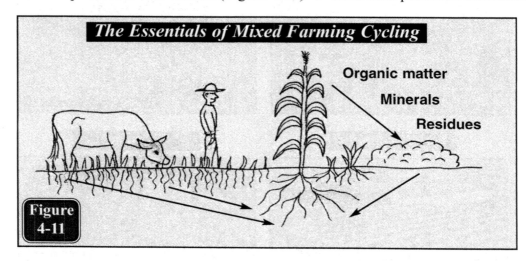

The Essentials of Mixed Farming Cycling

Organic matter

Minerals

Residues

Figure 4-11

ly for decades as the local cycle of soil to animal to man, and back to the soil again was followed religiously. The net annual export of nutrients from farms was minimal because crop yields were considerably less back in the 1800's and early 1900's. Erosion took its toll, but some farmers were astute enough to limit that as well.

Today, yields are much higher than in America's early years. Nutrient flow is massive from the soil of the country's breadbasket to population centers. Organic matter is being depleted continually as well, and erosion by water and wind are as great as ever. Only by artificially feeding plants with synthetic NPK, lime, and other manufactured nutrients are farmers able to keep producing crops while barely surviving economically. Some do not survive, and every year many sell out so we now have only about 1.8% of the population in commercial farming. Mixed farming is practiced much less today than only 50 years ago as livestock production has become largely centralized in huge feedlots, dairies, and pig and poultry factories. Farm animals have in many cases been separated from crop farms as smaller family farms disappear, so manure — so important for soil fertility maintenance — is not usually cycled back to the soil.

8. **Optimum soil biological activity.** It should be obvious by now that the key to fertile soil is a high organic matter level ... and the heart of organic matter activity is its teeming microbial community. We have shown how rhizosphere soils are tremendously more active in biological activity then are non-rhizosphere soils, and it is within this active root zone that most nutrient release occurs. Nearly all of that release is mediated by microorganisms and earthworms.

We have seen how critical it is to conform to nature's God-given laws in order to maximize soil fertility by enhancing natural biological processes. It is possible to work *with* — not *against* — nature's laws and build up the land to its optimum potential. Each location on earth has unique needs for its soils, and we are able to optimize that soil's production by balancing nutrients with manure and natural ground rock fertilizers, reducing or eliminating tillage, planting adapted crop varieties, and utilizing the land according to its capabilities.

Man was truly designed to have **dominion** [Hebrew *radah*, "subjugate"] over the fish, birds, and animals, and certainly the soil as well[25]. Dominion does not mean abuse, destroy, and pummel. It means treat with kindness and love, with wisdom and knowledge, with compassion and humility. Being placed over the creation has been a tremendous responsibility granted by the Creator, one that requires continual and judicious stewardship. Rather than be a selfish pariah to the soil, a parasite, mankind is to be a mutualistic leader in preserving and protecting that all-important six inches that stands between himself and starvation. Though today's economic and political system — patterned after Satan's designs — is hellbent on running farmers off the land, and making those that remain slaves of exploitative profit maximization, there is hope. Imposing an industrial ethic upon a biological system cannot work forever, for every law broken exacts its recompense. We have seen that profits from organic farming are as great as from conventional farming. Mixed farming still pays, especially for specialty and high value crops. Premiums are paid for organic produce because they truly are superior to those raised conventionally[26].

We live in a culture dominated by forces of both good and evil. Will we become mutualistic with the God who made all things good, and treat our soils with respect as we ought to treat our neighbors?

Bibliography

Preface

1. Proverbs 1:7.
2. Proverbs 9:10.
3. Genesis 1:26, King James Version with modifications. This version, sometimes with slight modifications for ease of reading, will be used throughout this text. The original word for *God* is *Elohim*, which means "Gods", the plural form of God.
4. Gesenius, W., Brown, F., Driver, S., and Briggs, C. 1907. The New Brown, Driver, and Briggs Hebrew and English Lexicon of the Old Testament. Reprinted in 1981 by Associated Publishers and Authors, Inc., Lafayette, Indiana.
5. Genesis 2:8,15.
6. See 4.
7. See 4.
8. Matthew 7:12.
9. Matthew 22:37-40.
10. Job 1:6-12; 2:1-7; Psalms 17:4; Isaiah 14:12; Matthew 4:1-11; John 12:31; 16:11; II Corinthians 4:4; Ephesians 2:2.
11. See the creation in Genesis 1: all things were made "very good"
12. Exodus 7 to 12.

Chapter I

1. Romans 8:18-20.
2. Genesis 1:4, 10, 12, 18, 21, 25, 31.
3. Job 1:6-12; Psalms 17:4; Isaiah 14:12; Matthew 4:1-11.
4. II Corinthians 4:4; Ephesians 2:2; 6:12; John 12:31; 14:30; 16:11.
5. Genesis 2:19-20. The lion and other vicious meat eaters were docile in the presence of Adam, not predacious as they are today.
6. Genesis 3:18; Revelation 21:4.
7. Genesis 1:31.
8. Guralnik, D. (editor). 1987. *Webster's New World Dictionary of the American Language*. Warner Books, Inc., New York, New York.
9. See 8.
10. Considine, D. (editor). 1976. *Van Nostrand's Scientific Encyclopedia*, Fifth Edition. Van Nostrand Reinhold Company, New York, New York.
11. Sylvia, D., Fuhrmann, J., Hartel, P., and Zuberer, D. (editors). 1998. *Principles and Applications of Soil Microbiology*. Prentice-Hall, Inc., Upper Saddle River, New Jersey.
12. MacQuitty, M. 1988. *Side By Side*. G. P. Putnam's Sons, New York, New York.
13. See 12.
14. Mai, W., and Lyon, H. 1975. *Pictorial Key to Genera of Plant-Parasitic Nematodes*, Fourth Edition, Revised. Cornell University Press, Ithaca, New York.

15. Adapted from a photo in Baker, K., and Snyder, W. (editors), 1965, *Ecology of Soil-Borne Plant pathogens, Prelude to Biological Control*, An International Symposium On Factors Determining the Behavior of Plant Pathogens in Soil, at the University of California, Berkeley, California, April 7-13, 1963, University of California Press, Berkeley, California.
16. Drawing by Paul Syltie.
17. Malachi 3:10.
18. Galations 6:9-10.
19. Proverbs 19:17.
20. Ecclesiastes 11:1.
21. II Corinthians 9:6,8.
22. Matthew 10:42.
23. Deuteronomy 15:8.
24. Proverbs 21:26.
25. Matthew 7:7-12.
26. Luke 6:27-31, 35, 38; see also Matthew 7:2 and Mark 4:24.

Chapter II

1. Drawing by Paul Syltie.
2. Notice Matthew 20:25-28, where it is stated that "... whoever will be chief among you, let him be your servant ...."
3. Lehninger, A. 1975. *Biochemistry*. Worth Publishers, Inc., New York, New York.
4. From Janick, J., Schery, R., Woods, F., and Rutton, V. 1969. *Plant Science*. W.H. Freeman and Company, San Francisco, California.
5. See 4.
6. Eames, A., and MacDaniels, L. 1947. *An Introduction to Plant Anatomy*. McGraw-Hill, New York, New York.
7. Adapted from Janick et al. (see 4).
8. See 1.
9. Hall, J., Flowers, T., and Roberts, R. 1974. *Plant Cell Structure and Metabolism*. Longman Group Limited, London, England.
10. See 4.
11. Adapted from 3 and 9.
12. Adapted from 3.
13. Jenny, H. 1941. *Factors of Soil Formation*. McGraw-Hill Book Company, New York, New York; Kellogg, C. 1936. The development and significance of the great soil groups of the United States. U.S.D.A. *Miscellaneous Publication 229*. U.S. Government Printing Office, Washington, D.C.
14. Donahue, R., Shickluna, J., and Robertson, L. 1971. *Soils, An Introduction to Soils and Plant Growth*, Third Edition. Prentice-Hall, Inc., Englewood Cliffs, New Jersey.
15. Buckman, H., and Brady, N. 1969. *The Nature and Properties of Soils,* Seventh Edition. The Macmillan Company, New York, New York.
16. Joffe, J., and Kunin, R. 1942. Mechanical separates and their fraction in the soil pro-

file: I. Variability in chemical composition and its pedogenic and agropedogenic implications. *Proceedings of the Soil Science Society of America* 7:187-193.

17. Foth, H., and Ellis, B. 1988. *Soil Fertility*. John Wiley and Sons, Inc., New York, New York.

18. Boh, H., McNeal, B., and O'Connor, G. 1985. *Soil Chemistry*. John Wiley and Sons, New York, New York

19. Failyer, G., et.al. 1908. The mineral composition of soil particles. U.S.D.A., Bureau of Soils, *Bulletin 54*. U.S. Government Printing Office, Washington, D.C.

20. Kabata-Pendias, A., and Pendias, H. 1992. *Trace Elements in Soil and Plants*. CRC Press, Boca Raton, Florida.

21. See 15.

22. Pearson, R., and Simonson, R. 1939. Organic phosphorus in seven Iowa profiles: distribution and amounts as compared to organic carbon and nitrogen. *Proceedings of the Soil Science Society of America* 4:162-167.

23. See 21 and many other soils and agronomic texts.

24 to 28. See 15.

29. Kinsey, N., and Walters, C. 1993. *Neal Kinsey's Hands-On Agronomy*. Acres U.S.A., Kansas City, Missouri.

30 to 37. See 15.

38. White, W., and Collins, D. (Editors). 1976. *The Fertilizer Handbook*. The Fertilizer Institute, Washington, D.C.; Russell, E. 1973. *Soil Conditions and Plant Growth*, Tenth Edition. Longman Group Limited, London, England.

39. See 20. The following table shows elemental interactions that are known to occur within plants. Kabata-Pendias and Pendias (see 20) list dozens more interactions of ele-

Interactions Among Major Elements and Trace Elements in Plants		
Major element	Antagonistic elements	Synergistic elements
Ca	Al, B, Ba, Be, Cd, Co, Cr, Cs, Cu, F, Fe, Li, Mn, Ni, Pb, Sr, and Zn	Cu, Mn, and Zn
Mg	Al, Be, Ba, Cr, Mn, F, Zn, Ni, Co, Cu, and Fe	Al, and Zn
P	Al, As, B, Be, Cd, Cr, Cu, Fe, Hg, Mo, Mn, Ni, Pb, Rb, Se, Si, Sr, and Zn	Al, B, Cu, F, Fe, Mo, Mn, and Zn
K	Al, B, Hg, Cd, Cr, F, Mo, Mn, Rb.	—
S	As, Ba, Fe, Mo, Pb, Se, and Zn	F, and Fe
N	B, F, Cu, and Mn	B, Cu, Fe, and Mo
Cl	Br, and I	—
Na	Mn	—
Si	B, and Mn	—

ments within the soil itself.

40. Charles, R. 1993. *The Mineral Content of Foods: Organic and Conventional Compared.* Master of Science Study. University of Bridgeport, Bridgeport, Connecticut.

41. See 40.

42. Bolliger, M. 1998. Food and health: what doctors and farmers share in common. *Eco-Farm and Garden.* Summer, 1998.

43. Pill, W. 1997. Filling the void. *American Nurseryman* 185 (12):30-37.

44. Gleick, J. 1985. Quiet clay revealed as vibrant and primal. *The New York Times*, New York, New York

45. Grim, R., 1953. *Clay Mineralogy.* McGraw-Hill Book Company, Inc., New York, New York.

46. See 14.

47. Arneman, H. 1964. *Intermediate Soils* (Soils 19) notes. University of Minnesota College of Agriculture, Forestry, and Home Economics, St. Paul, Minnesota.

48. Adapted from 15.

49. Adapted from 47.

50. Adapted from 47.

51. See 15.

52. See 1.

53. See 47.

54. See 21.

55 to 58. See 44.

59. Wilford, J. 1985. New finding lacks idea that life started in clay rather than sea. *The New York Times*, April 3, 1985.

60. Gesenius, W., Brown, F., Driver, S., and Briggs, C. 1907. The New Brown, Driver, and Briggs Hebrew and English Lexicon of the Old Testament. Reprinted in 1981 by Associated Publishers and Authors, Inc., Lafayette, Indiana.

61. See Lisle, H. 1994. *The Enlivened Rock Powders.* Acres U.S.A., Metairie, Louisiana.

62. Callahan, P. 1995. *Paramagnetism: Rediscovering Nature's Secret Force of Growth.* Acres U.S.A., Metairie, Louisiana.

63. See 62.

64. Data adapted from *Handbook of Chemistry and Physics*, 50th Edition. Chemical Rubber Company, Boca Raton, Florida.

65. See 62.

66. Syltie, P. 1998. A closer look at organic matter. *The Vital Earth News—Agricultural Edition* 3 (1):6-7.

67. Kononova, M. 1966. *Soil Organic Matter.* Pergamon Press, Inc., Oxford, England.

68. See 67.

69. Foster, R., Rovira, A., and Cock, T. 1983. *Ultrastructure of the Root-Soil Interface.* The American Phytopathological Society, St.Paul, Minnesota.

70. See 69.

71. See 67.

72. See 4, 15, and 66.

73. Bacterial rods, see 69. bacterial pseudomonads, see 69; fungal VA mycorrhiza, Paul, E., and Clark, F., 1989, *Soil Microbiology and Biochemistry*, Academic Press, Inc., New

York, New York; fungal VA arbuscule, see Paul and Clark, 1989; soil fungus, Killham, K., 1994, *Soil Ecology*, Cambridge University Press, Cambridge, England; protozoa, Killham, 1994; actinomycetes, Dindal, D., 1990, *Soil Biology Guide*, John Wiley and Sons, Inc., New York, New York; nematodes, Dropkin, V., 1989, *Introduction to Plant Nematology*, Second Edition, John Wiley and Sons, New York, New York; algae, Sylvia, D., Fuhrman, J., Hartel, P., and Zuberer, D., 1998, *Principles and Applications of Soil Microbiology*, Prentice-Hall, Inc., Saddle River, New Jersey; earthworms, photo by Paul Syltie; mites, Dindal, 1990; azotobacter, Paul and Clark, 1989; rhizobium, Sylvia, D., Fuhrman, J., Hartel, P., and Zuberer, D., 1998, *Principles and Applications of Soil Microbiology*, Prentice-Hall, Inc., Saddle River, New Jersey, from Perotto, S., Brewin, N., and Kannenberg, E., 1994, Cytological evidence for a host defense response that reduces cell and tissue invasion in pea nodules by lipopolysaccharide-defective mutants of *Rhizobium leguminosarum* strain 3841, *Molecular Plant Microbe Interactions* 7:99-112; tardigrades, Dindal, 1990.

74. Ingham, E. 1999. What do different plants need? SoilFoodWeb, Inc. *www.soilfood-web.com.*

75. See 15.

76. See Paul and Clark, 1989, 73.

77. See Paul and Clark, 1989, 73.

78. Campbell, C., Paul, E., Rennie, D.A, and McCellun, K.J. 1967. *Soil Science* 104:81, 217.

79. Ruke, R.V., 1969. *Soil Science* 107:318.

80. See 15.

81. Syltie, P.W. 1998. Soil structure. *The Vital Earth News—Agricultural Edition* 3 (2):4-5.

82. See 1.

83. See 15.

84. Foster, R.C. 1981. Polysaccharides in soil fabrics. *Science* 214:665-667.

85. See 15.

86. See 69.

87. From Metzger, l., et al. 1984. *Soil Science Society of America Journal* 48. In Syltie, P. 1998-1999. Soil structure, part 2, Lesson 8: 15-minute soils course *The Vital Earth News — Agricultural Edition* 4(1). Vital Earth Resources, Gladewater, Texas.

88. Syltie, P. 1998. Soil structure, Lesson 7: 15-minute soils course. *The Vital Earth News — Agricultural Edition* 3(2). Vital Earth Resources, Gladewater, Texas.

89 to 91. See 15

92. Adapted from 15.

93. See 66.

94. Adapted from 15

95. See 15.

96. Shinitzky, M., and Henkart, P. 1979. Fluidity of cell membranes — current concepts and trends. *International Review of Cytology* 60:121-147; Hall, J., Flowers, T., and Roberts, R. 1974. *Plant Cell Structure and Metabolism*. Longman Group Limited, London, England.

97. See 1.

98 and 99. See 15.

100. See 1.

101 and 102. See 15.

103 and 104. See 4.

105. See 1.

106. See 4.

107. Adapted from 15.

108. Adapted from 15.

109 and 110. See 15.

111. See 1.

112. See 15.

113. Adapted from 4 and 15.

114. See 4.

115. Matthew 5: 44-45.

116. See 15, as copied from U.S. Soil Conservation Service documents.

117. Adapted from 15.

118. See 1.

119. Photos by Paul Syltie: forest soil from eastern Texas; prairie soil from east-central Illinois; steppe soil from central Wyoming.

Chapter III

1. Smith, A. 1977. Microbial interactions in soils and healthy plant growth. *Australian Plants* 9:73.

2. Peirson, D., and Dumbroff, E. 1969. Demonstration of a complete Casparian strip in *Avena* and *Ipomoea* by a fluorescent staining technique. *Canadian Journal of Botany* 47:1869-1871.

3. Adapted from Menge, J., 1985, Mycorrhiza agriculture technologies, Chapter X in *Innovative Biological Technologies for Lesser Developed Countrie*s. Proceedings of a workshop, July, Office of Technology Assessment (OTA-BP-F-29), Washington, D.C.

4. Adapted from Epstein, E. 1972. *Mineral Nutrition of Plants: Principles and Perspectives*. John Wiley and Sons, Inc., New York, New York

5. See 4.

6. Hall, J., Flowers, T., and Roberts, R. 1974. *Plant Cell Structure and Metabolism*. Longman Group Limited, London, England. Nye, P. and Tinker, P. 1977. Solute movement in the soil root system. *Volume 4: Studies in Ecology*, University of California Press, Berkeley, California, U.S.A.

7. See 4.

8. Hanawalt, R., and Whittaker, R. 1977. Altitudinal gradients of nutrient supply to plant roots in mountain soils. *Soil Science* 123:85-96.

9. See 3, and Redhead, J. 1968. Mycorrhizal associations in some Nigeria forest trees. *Transactions of the British Mycology Society* 51:377-387.

10. Bieleski, R. 1973. Phosphate pools, phosphate transport, and phosphate availability. *Annual Review of Plant Physiology* 24:225-252.

11. Barber, S. 1984. *Soil Nutrient Bioavailability, A Mechanistic Approach*. John Wiley and Sons, New York, New York. Foster, R. 1981. The ultrastructure and histochemistry of the rhizosphere. *New Phytology* 89:263-273. Keltjens, W. 1982. Nitrogen metabolism and K-recirculation in plants. In *Plant Nutrition 1982*, Proceedings of the Ninth International Plant Nutrition Colloquium (edited by A. Scaife), 283-287. Warwick University, England. Commonwealth Agricultural Bureau, Farnhamhouse, England.

12. Anghinoni, I., and Barber, S. 1980. Predicting the most efficient phosphorus placement for corn. *Soil Science Society of America Journal* 44:1016-1020. Foy, C. 1974. Effects of aluminum on plant growth. In *The Plant Root and Its Environment* (edited by E. Carson), 601-642. University Press of Virginia, Charlottesville, Virginia, U.S.A. Troughton, A. 1980. Environmental effects upon root-shoot relationships. In *Environment and Root Behavior* (edited by D. Sen), pp. 25-41. Geobios International, Jodhpur, India. Walker, J. 1969. One-degree increments in soil temperatures affect maize seedling behavior. *Soil Science Society of America Proceedings* 33:729-736.

13. See 8.

14. See 8.

15. Melsted, S. 1970. Absorption of nutrient elements. Mimeo, Department of Agronomy, University of Illinois, Urbana, Illinois, U.S.A.

16. Bieleski, R. 1973. Phosphate pools, phosphate transport, and phosphate availability. *Annual Review of Plant Physiology* 24:225-252.

17. Rhodes, L., and Gerdemann, J. 1975. Phosphate uptake zones of mycorrhizal and nonmycorrhizal onions. *New Phytology* 75:555-561.

18. Mengel, K., Grimme, H., and Nemeth, K. 1969. Potentielle und effective Ver fugbarkeit von Pflanzennahrstoffen in Boden. *Landwirtschaftliche Forschung* 23:79-91.

19. Jenny, H., and Grossenbacher, K. 1962. Root soil boundary zones. *California Agriculture* 16:7. Rovira, A. 1965. Plant root exudates and their influence upon soil microorganisms. In *Ecology of Soil-Borne Plant Pathogens, Prelude to Biological Control* (edited by K. Baker and W. Snyder), pp. 170-186. University of California Press, Berkeley, California. Scott, F. 1965. The anatomy of roots. In *Ecology of Soil-Borne Plant Pathogens, Prelude to Biological Control* (edited by K. Baker and W. Snyder), pp. 145-153. University of California Press; Berkeley, California.

20. See 15.

21. Barber, 1984 (see 11).

22. See 3.

23. Rhodes, L., and Gerdemann, J. 1975. Phosphate uptake zones of mycorrhizal and nonmycorrhizal onions. *New Phytology* 75:555-561.

24. Finch, P., Hayes, M., and Stacy, M. 1971. In *Soil Biochemistry*, Volume 2 (edited by A.McLaren and J. Skujins). Dekker; New York, New York.

25. Ramamoorthy, S., and Leppard, G. 1976. A discrete physiological role for organic colloid debris in soil. *Naturwissenschaften* 63:579.

26. See 25.

27. Nielsen, N. 1976. A transport kinetic concept for ion uptake by plants. 3. Test of the concept by results from water culture and pot experiments. *Plant and Soil* 45:659-677.

28. Nye, P. and Tinker, P. 1977. *Solute Movement in the Soil Root System*. Volume 4: Studies in Ecology, University of California Press, Berkeley, California, U.S.A.

29. Russell, E. 1973. *Soil Conditions and Plant Growth*, Tenth Edition. Longman Group Limited, London, England. Leavitt, J. 1974. *Introduction to Plant Physiology*, Second Edition. The C.V. Mosby Company, St. Louis, Missouri.

30. Barber, S. 1962. A diffusion and mass-flow concept of soil nutrient availability. *Soil Science* 93:29-49. Brewster, J., and Tinker, P. 1972. Nutrient flow rates into roots. *Soil and Fertilizer* 35:355-359.

31. Mosse, B. 1973. Advances in the study of vesicular-arbuscular mycorrhiza. *Annual Review of Phytophathology* 1:171-196. Possingham, J., and Obbink, J. 1971. Endotrophic mycorrhiza and the nutrition of grape vines. *Vitis* 10:120-130. Ross, J., and Harper, J. 1970. Effect of endogone mycorrhiza on soybean yields. *Phytopathology* 60:1552-1556.

32. LaRue, J., McClellan, W., and Peacock, W. 1975. Mycorrhizal fungi and peach nursery nutrition. *California Agriculture* 29:6-7. Mosse, B. 1973. Advances in the study of vesicular-arbuscular mycorrhiza. *Annual Review of Phytophathology* 1:171-196.

33. Ross, J. 1971. Effect of phosphate fertilization on yield of mycorrhizal and nonmycorrhizal soybeans. *Phytopathology* 61:1400-1403.

34. Mosse, B. 1957. Growth and chemical composition of mycorrhizal and non-mycorrhizal apples. *Nature* 179:922-924.

35. Holevas, C. 1966. The effect of a vesicular-arbuscular mycorrhiza on the uptake of soil phosphorus by strawberry (*Fragaria* sp. var. *Cambridge Favourite*). *Journal of Horticulture Science* 41:57-64. Mosse, B. 1973 (see 31). Scannerini, S., Bonfante, P., and Fontana, A. 1975. Fine structure of the host-parasite interface in endotrophic mycorrhiza of tobacco. In *Endomycorrhizas* (edited by F. Sanders, B. Mosse, and P. Tinker), pp. 325-334. Academic Press, New York, New York.

36. Mosse, 1973 (see 31). See 33.

37. Mosse, 1973 (see 31).

38. Bowen, G. 1969. Nutrient status effects on loss of amides and amino acids from pine roots. *Plant and Soil* 30:139-143. Rovira, A. 1965. Plant root exudates and their influence upon soil microorganisms. In *Ecology of Soil-Borne Plant Pathogens, Prelude to Biological Control* (edited by K. Baker and W. Snyder), pp. 170-186. University of California Press, Berkeley, California

39. Redhead, J. 1968. Mycorrhizal associations in some Nigeria forest trees. *Transactions of the British Mycology Society* 51:377-387.

40. Backman, P., and Clark, E. 1977. Effect of carbofuran and other pesticides on vesicular-arbuscular mycorrhizae in peanuts. *Nematropica* 7:13-15. Jalali, B., and Domsch, K. 1975. Effect of systemic fungi toxicants on the development of endotrophic mycorrhiza. In *Endomycorrhizas* (edited by F. Sanders, B. Mosse, and P. Tinker), pp. 619-626. Academic Press, New York, New York. Sutton, J., and Sheppard, B. 1976. Aggregation of sand-dune soil by endomycorrhizal fungi. *Canadian Journal of Botany* 54:326-330.

41. See 3.

42. Menge, J., Davis, R., Johnson, E., and Zentmeyer, C. 1978. Mycorrhizal fungi increase growth and reduce transplant injury in avacados. *California Agriculture* 32:6-7. Menge, J., Gerdemann, J., and Lembright, H. 1975. Mycorrhizal fungi and citrus. *The Citrus Industry* 61:16-18.

43. See 1.

44. Rovira, A., and Davey, C. 1974. Biology of the rhizosphere. In *The Plant Root and*

Its Environment (edited by E. Carson), pp.153-204. University Press of Virginia, Charlottesville, Virginia

45. Rovira, A. 1965. Plant root exudates and their influence upon soil microorganisms. In *Ecology of Soil-Borne Plant Pathogens, Prelude to Biological Control* (edited by K. Baker and W. Snyder), pp. 170-186. University of California Press, Berkeley, California

46. Dobereiner, J., and Day, J. 1975. Nitrogen fixation in the rhizosphere of tropical grasses. In *Nitrogen Fixation by Free Living Microorganisms* (edited by W. Steward), pp. 39-56. Cambridge University Press, New York, New York.

47. Katznelson, H. 1965. Nature and importance of the rhizosphere. In *Ecology of Soil-Borne Plant Pathogens, Prelude to Biological Control* (edited by K. Baker and W. Snyder), pp. 187-209. University of California Press, Berkeley, California. Vrany, J., Vancura, V., and Macura, J. 1962. The effect of foliar applications of some readily metabolized substances, growth regulators, and antibiotics on rhizosphere microflora. *Folia Microbiologica*, Prague 7:61-70.

49. Fisher, J., Hanson, D., and Hodges, T. 1970. Correlation between ion fluxes and ion-stimulated adenosine triphosphate activity of plant roots. *Plant Physiology* 46:812-814. Hodges, T. 1973. Ion absorption by plant roots. *Advances in Agronomy* 25:163-207.

50. Hewitt, E., and Smith, T. 1975. *Plant Mineral Nutrition*. The English University Press, Ltd., London, England.

51. Barber, 1984 (see 11). See 4 and 29.

52. Ashton, F., and Crafts, A. 1981. *Mode of Action of Herbicides*. John Wiley and Sons, New York, New York. Brady, N. 1974. *The Nature and Properties of Soils*, Eighth Edition. Macmillan Publishing Company, Inc., New York, New York.

53. Isensee, Q., Jones, G., and Turner, B. 1971. Root absorption and translocation of picloram by oats and soybean. *Weed Science* 19:727-731.

54. Donaldson, T., Bayer, D., and Leonard, O. 1973. Absorption of 2, 4-dimethylurea (monuron) by barley roots. *Plant Physiology* 52:638-645.

55. Nishizawa, N., and Mori, S. 1977. Invagination of plasmalemma: its role in the absorption of macromolecules in rice roots. *Plant and Cell Physiology* 18:767-782.

56. Mori, S., and Nishizawa, N. 1976. Endocytosis in plant cells. *Abstracts of the 1976 Meeting, Society of the Science of Soil and Manure, Japan* 22: Part II, 67. Mori, S., and Uchino, H. 1976. Criticism of the mineral nutrition theory, III. Growth features of rice water-cultured with organic nitrogens. *Abstracts of the 1976 Meeting, Society of the Science of Soil and Manure, Japan* 22:67. Mori, S., and Yumoto, S. 1976. The mechanism of protein uptake in plant cells: possible involvement of phagocytosis. *Journal of Cell Biology*, Abstracts of the First International Congress on Cell Biology 70:281.

57. Tisdale, S., and Nelson, W. 1966. *Soil Fertility and Fertilizers*. The Macmillan Company, New York, New York.

58. Brady, 1974 (see 52). See 29.

59. Danielli, J., and Davson, H. 1935. A contribution to the theory of permeability of thin films. *Journal of Cellular Composition and Comparative Physiology* 5:495-508. See 6.

60. See 59.

61. Benson, A. 1966. On the orientation of lipids in chloroplast and cell membranes. *Journal of the American Oil Chemical Society* 48, 265. Frey-Wyssling, A., and Muhlethaler, K. 1965. *Ultrastructural Plant Cytology*. Elsevier, Amsterdam, Netherlands.

Lenard, J., and Singer, S. 1966. Protein conformation in cell membrane preparations as studied by optical-rotatory dispersion and circular dichroism. *Proceedings of the National Academy of Science* (U.S.A.), 56:1828-1835. Lucy, J. 1968. Ultrastructure of membranes; micellar organization. *British Medical Bulletin* 24:127-129. Muhlethaler, K., Moor, H., and Szarkowski, J. 1966. The ultra structure of the choloroplast lamella *(Spinacea oleracea)*. *Planta* 67:305-323. Robertson, J. 1967. The organization of cellular membranes. In *Molecular Organization and Biological Function* (edited by J. Allen). Harper and Row, New York, New York.

62. Lenard and Singer, 1966 (see 61).

63. See Hall et al., 1974 (see 6).

64 Duysen, M. 1979. North Dakota State University, Fargo, North Dakota, U.S.A. Personal communication.

65. Zeevi, A., and Margalit, R. 1985. Selective transport of lithium across lipid bilayer membranes mediated by an ionophore of novel design (ETH 1644). *Journal of Membrane Biology* 86:61-68.

66. Lehninger, A. 1977. *Biochemistry*. Worth Publishers, Inc., New York, New York.

67. Barber, S. 1984. *Soil Nutrient Bioavailability, a Mechanistic Approach*. John Wiley and Sons, New York, New York; Levitt, J. 1974. *Introduction to Plant Physiology*. The C.V. Mosby Company, St. Louis, Missouri.

68. Barber, 1984 (see 11). See 66.

69. See 4 and 66.

70. See 66.

71. Hall et al., 1974 (see 6).

72. See 4.

73. See 28.

74. See 4. Truter, M.R. 1975. Coordination chemistry of alkali metal cations. *Annual Report, Rothamsted Experimental Station, 1974*, 165-169.

75. See 66.

76. Epstein, E. 1953. Ion absorption by plant roots. In *Proceedings of the Fourth Annual Oak Ridge Summer Symposium* (edited by C. Comar and S. Hood). Oak Ridge, Tennessee.

77. Glass, A. 1976. Regulation of potassium absorption in barley roots; an allosteric model. *Plant Physiology* 58:33-37.

78. Barber, 1984 (see 11).

79. See 4.

80. Laties, G. 1969. Dual mechanism of salt uptake in relation to compartmentation and long-distance transport. *Annual Revue of Plant Physiology* 20:89-116.

81. See 55.

82. See 28.

83. Brachet, J. 1956. The mode of action of ribonuclease on living root tip cells. *Biochimica et Biophysica Acta* 19:583.

84. McLaren, A., Jensen, W., and Jacobson, L. 1960. Absorption of enzymes and other proteins by barley roots. *Plant Physiology* 35:544-556.

85. Coulomb, C. 1973. Diversite' des corps multive'siculaires et notion d'heterophagae dan le m'riste'me radiculair de scorsone're *(Scorzoners hispanica)*. *Journal Microscopie* 16:345-360. Mahlberg, P., Turner, F., Walkinshaw, C., and Venketeswaran, S. 1974.

Ultrastructural studies of plasma membrane related secondary vacuoles in cultured cells. *American Journal of Botany* 61:730-738. Robards, A., and Robb, M. 1974. The entry of ions and molecules into roots: an investigation using electron-opaque tracers. *Planta* 120:1-12. Sung, Z., and McLaren, A. 1975. Entry of a basic protein, lysozyme, through the region of elongation in roots of *Iasione montana. Plant and Cell Physiology* 16:455-464.

86. Robards, A., and Robb, M. 1974. The entry of ions and molecules into roots: an investigation using electron-opaque tracers. *Planta* 120:1-12.

87. Hall et al., 1974 (see 6).

88. Clarkson, D. 1974. *Ion Transport and Cell Structure in Plants*. McGraw Hill Book Company, Inc., London.

89. See 28.

90. Drawing by Paul W. Syltie. See Buckman, H., and Brady, N. 1969. *The Nature and Properties of Soils,* Seventh Edition. The Macmillan Company, New York, New York. Waksman, S. 1948. *Humus.* William and Wilkins, Baltiman, Maryland.

91. Paul, E., and Clark, F. 1989. *Soil Microbiology and Biochemistry*. Academic Press, Inc, New York, New, York.

92. Syltie, P. 1985. Effects of very small amounts of highly active biological substances on plant growth. *Biological Agriculture and Horticulture* 2: 245-269.

93. Buckman and Brady, 1969 (see 90). Stevenson, F. 1985. *Cycles of Soil: Carbon, Nitrogen, Phosphorus, Sulfur, Micronutrients*. John Wiley and Sons, New York, New York.

94. Photograph from Mishustin, E. and Shil'nikova, V. 1971. *Biological fixation of Atmospheric Nitrogen*. The Pennsylvania State University Press, University Park, Pennsylvania.

95. Stevenson, 1985 (see 93).

96. Rateaver, B., and Rateaver, G. 1993. *Organic Method Primer Update*, Special Edition. The Rateavers, San Diego, California.

97. Ingham, E. 1999. The soil foodweb. *www.soilfoodweb.com.* Soil FoodWeb Incorporated, Corvallis, Oregon.

98. Drawing by Greg and Minette Smith.

99. Ingham, E. 1999. What do different plants need. *www.soilfoodweb.com.* SoilFoodweb Incorporated, Corvallis, Oregon.

100. See 98.

101. Anonymous. 1981. Spit spurs growth. *Science Digest* 89:25.

102. Genesis 1:30.

103. Russell, E., and Russell, F. 1950. *Soil Conditions and Plant Growth..* Longmans and Green, New York, New York.

104. Vimmerstedt, J. 1969. Earthworm speed leaf decay on spoilbanks. *Ohio Report* 54 (1):3-5.

105. Buckman and Brady, 1969 (see 90).

106. Drawing by Paul W. Syltie.

107. Adapted from Buckman and Brady, 1969 (see 90).

108. Technical Committee On Soil Survey. 1960. Soils of the North Central Region of the United States. *North Central Regional Publication No. 76.* Agricultural Experiment

Station, University of Wisconsin, Madison, Wisconsin.

109. Frank, L. 1990. *The Big Splash*. Carol Publishing Group, New York, New York.

110. Foster, R., Rovira, A., and Cock, R. 1983. *Ultrastructure of the Root-Soil Interface*. The American Phytopathological Society, St. Paul, Minnesota.

111. Baker, K., and Cook, R. 1974. *Biological Control of Plant Pathogens*, Reprint Edition (1982). American Phytopathological Society, St. Paul, Minnesota.

112. See 110.

113. Russell, R. 1977. Plant Root Systems: *Their Function and Interaction with the Soil*. McGraw-Hill Book Company, London, England.

114. Trouse, A., Jr. 1988. *Crop Root Capabilities*. National Soil Dynamics Laboratory, Agricultural Research Service, United States Department of Agriculture, Auburn, Alabama.

115. Miller, A., and Gow, N. 1989. Correlation between profile of ion-current circulation and root development. *Physiol. Plant* 75:102-108

116. See 106.

117. Adapted from Lynch, J. (editor). 1990. *The Rhizosphere*. John Wiley and Sons, New York, New York.

118 to 120. See 114.

121. Powell, C., and Bagyaraj, D. (editors). 1984. *VA Mycorrhiza*. CRC Press, Boca Raton, Florida.

122 to 124. See 110.

125. Adapted from 110, from a drawing by G. Rinder and R. Schuster.

126. See 106.

127. See 1. Whipps, J., and Lynch, J. 1985. Energy losses by the plant in rhizodeposition. *Plant Products and the New Technology, Annual Proceedings of the Phytochemical Society of Europe* (edited by U. Fuller and J. Callon) 26:59-71.

128. Starkey, R. 1958. Interrelations between microorganisms and plant roots in the rhizosphere. *Bacteriological Review* 22:154-172; Bokhari, U., Coleman, D., and Rubink, A. 1979. Chemistry of root exudates and rhizosphere soils of prairie plants. *Canadian Journal of Botany* 57:1473-1477.

129. Linderman, R. 1995. Personal communication. Oregon State University, Corvallis, Oregon.

130. See 110.

131. Callahan, P. 1994. *Exploring the Spectrum*. Acres U.S.A., Kansas City, Missouri.

132. See 131.

133. Callahan, P. 1975. *Tuning in to Nature*. The Devin-Adair Company, Old Greenwich, Connecticut.

134. See 133.

135. See 106.

136. Studies have shown that solutes can flow into and out of cells through microtubules, opening at the plasmodesmata of cells of all types. See 6.

137. See 98.

138. See 6.

139. Harley, J., and Smith, S. 1983. *Mycorrhizal Symbiosis*. Academic Press, New York, New York.

140. See 91.

141 to 143. See 3.

144. Read, D. The role of the mycorrhizal symbiosis in the nutrition of plant communities. In *Ecophysiology of Ectomycorrhizae of Forest Trees*. 1991 Marcus Wallenberg Prize Symposium, September 27. The Marcus Wallenberg Foundation, Falun, Sweden.

145 to 147. See 3.

148. Lynch, J. (editor). 1990. *The Rhizosphere*. John Wiley and Sons, New York, New York.

149. Photo by Paul Syltie.

150. Sylvia, D., Fuhrman, J., Hartel, P., and Zuberer, D. 1998. *Principles and Applications of Soil Microbiology*. Prentice-Hall, Inc., Saddle River, New Jersey.

151. Killham, K. 1994. *Soil Ecology*. Cambridge University Press, Cambridge, England.

152. See 151.

153. See 106.

154. Janick, J., Schery, R., Woods, F., and Rutton, V. 1969. *Plant Science*. W.H. Freeman Company, San Francisco, California.

155. See 148.

156. Balandreau, J., Ducerf, P., Homad-Fares, I., Weinhard, P., Rinaudo, G., Millier, C., and Dommergues, Y. 1978. Limiting factors in grass nitrogen fixation. In *Limitations and Potentials for Biological Nitrogen Fixation in the Tropics*, p. 275-302. (edited by J. Dobereiner, R. Burris, and A. Hollander). Plenum Press, New York, New York.

157. See 148.

158. Paul, E., and Clark, F. 1989. *Soil Microbiology and Biochemistry*. Academic Press, Inc., New York, New York.

159. See 94.

160. Roviera, A. 1965. Plant root exudates and their influence upon soil micro-organisms. In *Ecology of Soil-Borne Plant Pathogens, Prelude to Biological Control* (edited by K. Baker and W. Snyder). University of California Press, Berkeley, California. Epanchinov, A. 1982. Participation of vitamins in the nitrogen nutrition of plants. *Chemical Abstracts* 99, no. 37441. See 47. Waksman, S. 1952. *Soil Microbiology*. John Wiley and Sons, New York, New York. See 29. Avakyan, Z., and Afrikyan, F. 1981. Vitamin B_{12} in soils. *Biologicheskii Zhurnal Armenii* 34:253-258. Williams, S. 1982. Are antibiotics produced in soil? *Pedobiologia* 23:427-435. Kononova, M. 1966. *Soil Organic Matter*. Pergamon Press, Oxford, England. Khalafallah, M., Saber, M., and Abd-El-Maksoud, H. 1982. Influence of phosphate dissolving bacteria on the efficiency of superphosphate in a calcareous soil cultivated with *Vicia faba*. *Zeitschrift fuer Pflanzenernaehrung and Bodenkunde* 145:455-459.

161. Smith, A. 1976. Ethylene production by bacteria in reduced microsites in soils and some implications to agriculture. *Soil Biology and Biochemistry* 8:293-298. Smith, A., Milham, P., and Morrison, W. 1978. Soil ethylene production specifically triggered by ferrous iron. In *Microbial Ecology* (edited by M. Loutit and J. Miles). Springer-Verlag, Berlin, Germany.

162. See 1.

163. Smith et al., 1978 (see 161)

164. See 1.

165. See 106.

166. See 1.

167 and 168. See 106.

169. Anonymous. 1999. Ethylene biosynthesis triggers plant disease mechanism. *Acres U.S.A.* 29 (9):4. Syltie, P. 1987. *How Agrispon and Sincocin-AG Work.* Appropriate Technology Ltd., Dallas, Texas.

170. Barber, S. 1984. *Soil Nutrient Bioavailability.* John Wiley and Sons, New York, New York.

171 and 172. See 106.

173. Stewart, J. 1980-1981. The importance of P cycling and organic P in soils. *Better Crops with Plant Food*, Winter, p. 16-19.

174. See 114.

175. Drawings by Paul Syltie except for the upper right and lower right figures (see 150). The soil ped photograph is from 110.

176. See 97.

177. Tisdall, J., and Oades, J. 1982. Organic matter and water stable aggregates in soils. *Journal of Soil Science* 33:141-163.

178. Photo by Al Trouse, Jr.

179. See 97.

180. See 155.

181. Sasser, J., and Freckman, D. 1986. A world perspective on nematology: the role of the society. *Society of Nematologists, Silver Jubilee Meeting.* August 17-22, 1986. Orlando, Florida.

182. Cowling, E., and Horsfall, J. 1980. Prologue: how plants defend themselves. In *Plant Disease, An Advanced Treatise*, Volume V (edited by J. Horsfall and E. Cowling). Academic Press, New York, New York.

183. Syltie, P. 1998. Immunize plants? Nature does it; why can't we? *The Vital Earth News, Horticultural Edition* V (2):2.

184. Angier, N. 1992. Plants defy microbes with immune defense and self-mutilation. *The New York Times*, August 18.

185. Sasser, J., and Carter, C. (editors). 1985. *An Advanced Treatise On* Meloidogyne, *Volume I: Biology and Control.* North Carolina State University Graphics, Raleigh, North Carolina.

186. Uhlenbroek, J., and Bijloo, J. 1958. Investigations on nematicides. I. Isolation and structure of a nematicidal principle occurring in *Tagetes* roots. *Recl. Trav. Chim. Pays-Bas Belgium* 77:1004-1009.

187. Allen, E., and Feldmesser, J. 1971. Nematicidal activity of a-chaconine: effect of hydrogen ion concentration. *Journal of Nematology* 3:58-61. Kogiso, S., Wada, K., and Munakata, K. 1976. Isolation of nematicidal polycetylenes from *Carthamus tinctorius* L. *Agricultural and Biological Chemistry* 40:2085-2089. Scheffer, F., Kickuth, R., and Visser, J. 1962. Die Wurzelausscheidungen von *Eragostis curvula* (Schrad.) Nees und ihr Einfluss auf Wurzeknoten-nematoden. *Z. Pflanzenernaehr. Dueng. Bodenk. D.* 98:114-120.

188. Day, S. 1993. A shot in the arm for plants. *New Scientist*, January.

189. See 155.

190. See 110.

191 to 193. See 155.

!94. See 181.

195. Dropkin, V. *Introduction to Plant Nematology*. John Wiley and Sons, Inc., New York, New York.

196. Adapted from a photo in Baker, K., and Snyder, W. (editors), 1965, *Ecology of Soil-Borne Plant pathogens, Prelude to Biological Control*, An International Symposium On Factors Determining the Behavior of Plant Pathogens in Soil, at the University of California, Berkeley, California, April 7-13, 1963, University of California Press, Berkeley, California.

197. Oteifa, B. 1993. Personal communication.

198. Thomashow, L., and Weller, D. 1988. Role of a phenazine antibiotic from *Pseudomonas fluorescens* in biological control of *Gaeumannomyces graminis* var. *tritici Journal of Bacteriology* 170:3497-3508.

199. See 155.

200. Liu, L., Kloepper, J., and Tuzun, S. 1995. Introduction of systemic resistance in cucumber against Fusarium wilt by plant growth promoting rhizobacteria. *Phytopathology* 85:843-847.

201. Oostendorp, M., Dickson, D., and Mitchell, D. 1991. Population development of *Pasteuria penetrans* on *Meloidogyne arenaria*. *Journal of Nematology* 23:58-64.

202. See 106.

203. Adapted from 66.

204. This section is based on research performed by Francis Chaboussou, and recorded in *Les Plantes Malades des Pesticides — Bases Nouvelles d'une Prevention Contre Maladies et Parasite*s [*Plants Made Sick by Pesticides — New Basis for the Prevention of Disease and Pests*]. 1980. Debard, Paris, France. See also this subject in Lutzenberger, J. 1984. Crops and pests: are poisons the answer? *The Ecologist* 14(2).

205. See 204.

206 to 210. See 106.

211. Matthew 7:12; Luke 6:31.

212. Leviticus 19:18; Matthew 22:39; Mark 12:31; Luke 10:27; Romans 13:9; Galations 5:14; James 2:8.

213. Matthew 6:3.

214. Matthew 6:1-6.

215. See the entire section in Matthew 25:31-40 of the sheep at God's right hand who performed benevolent deeds for others, and in the process were praised for having served God Himself.

216. Simard, S., Perry, D., Jones, M., Myvold, D., Durall, D., and Molina, R. 1977. Net transfer of carbon between ectomycorrhizal tree species in the field. *Nature* 338, 7 August: 579-582.

217. Finlay, R., and Read, D. 1986. The structure and function of the vegetative mycelium of ectomycorrhizal plants. II. The uptake and distribution of phosphorus by mycelial strands interconnecting host plants. *New Phytology* 103: 157-165.

218. Finlay, R., and Read, D. 1986. The structure and function of the vegetative myceli-

um of ectomycorrhizal plants. I. Translocation of ^{14}C-labeled carbon between plants interconnected by a common mycelium. *New Phytology* 103: 143-156.

219. Simard, S., Jones, M., and Durall, D. 2001. Carbon and nutrient fluxes within and between mycorrhizal plants. In *Mycorrhizal Ecology*, edited by M.van der Heijden and I. Sanders. Springer-Verlag, Heidelberg, Germany.

220. See 219.

221. Read, D. Mycorrhizal fungi in natural and semi-natural plant communities. In *Ecophysiology of Ectomycorrhizae of Forest Trees: The Marcus Wallenburg Foundation Symposia Proceedings 7*. 1991 Marcus Wallenburg Prize Symposium, Stockholm, Sweden, September 27. The Marcus Wallenburg Foundation, Falun, Sweden.

222. See 219.

223. See 219. Jones, M., Durall, D., Harniman, S., Classen, D., and Simard, S. 1998. Ectomycorrhizal diversity of paper birch and Douglas fir seedlings grown in single-species and mixed plots in the ICH Zone of southern British Columbia. *Extension Note 19*. British Columbia Ministry of Forests Research Program, Kamloops, British Columbia, Canada.

224. Note Leviticus 25:39-43; Deuteronomy 15:7-11; Acts 20:35; Romans 15:1; Ephesians 4:28; I Thessalonians 5:14.

225. See 219.

226. See 219.

227. Baldwin, I., and Schultz, J. 1983. Rapid changes in tree leaf chemistry induced by damage: evidence for communication between plants. *Science* 221: 277-279 (15 July).

228. Kiessling, L., and Gestwicki, J. 2002. Bacteria mimic cell phone system to communicate. *Nature*, January 3. In www.rense.com, January 3, 2002.

229. France', R. 1923. *The Love Life of Plants*. A. and C. Boni, New York, New York.

230. See 229.

231. Tompkins, P., and Bird, C. 1973. *The Secret Life of Plants*. Harper Colophon Books, New York, New York.

232. See 231.

233. See 231.

234. See 231.

235. Callahan, P. 1975. *Tuning In to Nature*. The Devin-Adair Company, Old Greenwich, Connecticut.

236. See 235.

237. Personal communications with Robert Linderman and Elaine Ingham, Oregon State University, Corvallis, Oregon.

Chapter IV

1. Fukuoka, M. 1978. *The One-Straw Revolution*. Rodale Press, Emmaus, Pennsylvania.

2. Luke 6:38.

3. Tilman, D. 1998. The greening of the Green Revolution. *Nature*: November 19. Drinkwater, L., Wagner, P., and Sarrantonio, M. 1998. Legume-based cropping systems have reduced carbon and nitrogen losses. *Nature*: November 19.

4. See 3.

5. Faulkner, E. 1943. *Plowman's Folly*. University of Oklahoma Press, Norman, Oklahoma.

6. Donahue, R., Shickluna, J., and Robertson, L. 1971. *Soils, An Introduction to Soils and Plant Growth*, Third Edition. Prentice-Hall, Inc., Engelwood Cliffs, New Jersey.

7. Agricultural Research Department. 1974. *World Crisis in Agriculture*. Ambassador College Press, Pasadena, California; photo from Donahue, R., Shickluna, J., and Robertson, L. 1971. *Soils, An Introduction to Soils and Plant Growth*, Third Edition. Prentice-Hall, Inc., Englewood Cliffs, New Jersey.

8. Bennett, H. 1947. *Elements of Soil Conservation*, First Edition. McGraw-Hill Book Company, Inc., New York, New York.

9. Gesenius, W., Brown, F., Driver, S., and Briggs, C. 1907. The New Brown, Driver, and Briggs Hebrew and English Lexicon of the Old Testament. Reprinted in 1981 by Associated Publishers and Authors, Inc., Lafayette, Indiana. These words are used in Genesis 2:15; 3:23; 4:2,12; II Samuel 9:10; I Chronicles 27:26; Nehemiah 10:37; Proverbs 28:19; Jeremiah 27:11; Ezekiel 36:9, 24.

10. Gesenius et al., 1907 (see 9).

11. The command is found in Genesis 2:15.

12. Genesis 3:19.

13. Genesis 2:16.

14. Genesis 1:29.

15. Genesis 2:17.

16. Genesis 3: 17-19.

17. See 9.

18. Hodges, H. 1970. *Technology in the Ancient World*. Barnes and Noble Books, New York, New York.

19. Albrecht, W. 1980. *The Albrecht Papers, Volume I.* Acres U.S.A., Raytown, Missouri.

20. Bowman, G. 1992. Grass in alfalfa baffles bugs as it builds soil and suppresses weeds. *The New Farm*, May/June. Spurgeon, D. 1980. Fielding higher yields. *Next*, November/December.

21. See 6.

22. Figure 4-6: Committee On the Role of Alternative Farming Methods in Modern Production Agriculture. 1989. *Alternative Agriculture*. National Research Council. National Academy Press, Washington. D.C.; Figure 4-7: Kohnke, H., and Bertrand, A. 1959. *Soil Conservation*. McGraw-Hill Book Company, Inc., New York, New York; Figure 4-8: Cook, R. 1962. *Soil Management for Conservation and Production*. John Wiley and Sons, Inc., New York, New York; Figure 4-9: see Kohnke and Bertrand, 1959.

23. Syltie, P. 1981. *The New Eden: Millennial Agriculture*. Triumph Publishing Company, Altadena, California.

24. See 106.

25. For a review of the issue of how soils relate to food quality, note the following articles:

Bolliger, M. 1998. Food and health: what doctors and farmers share in common. *Eco-Farm and Garden*. Summer.

Smith, B. 1993. Organic foods vs. supermarket food: element levels. *The Journal of*

Applied Nutrition 45 (1).

Beasley, J., and Swift, T. 1989. *The Kellogg Report: The Impact of Nutrition, Environment, and Lifestyle On the Health of Americans*. Institute of Health, Policy and Practice, The Bard College Center. Bard College, Annondale-on-Hudson, New York.

Bourn, D. *The Nutritional Value of Organically and Conventionally Grown Food — Is There a Difference?* Department of Consumer Services, University of Otago, Dunedin, New Zealand.

Charles, R. 1993. *The Mineral Content of Foods: Organic and Conventional Compared*. Masters of Science Study, University of Bridgeport, Bridgeport, Connecticut.

Eggert, F., and Kahrman, C. 1984. Response of three vegetable crops to organic and inorganic nutrient sources. In *Organic Farming: Current Technology and Its Role in Sustainable Agriculture*. American Society of Agronomy, Madison, Wisconsin.

Plochberger, K. 1989. Feeding experiments: a criterion for quality estimation of biologically and conventionally produced foods. *Agriculture, Ecosystems, and Environment* 27:419-428.

Schuphan, W. 1974. Nutritional value of crops as influenced by organic and inorganic fertilizer treatments. *Qualitas Plantarum: Plant Foods for Human. Nutrition.* 23 (4):333-358.

Staiger, D. 1988. The nutritional value of foods from conventional and biodynamic agriculture. *IFOAM Bulletin* 4, March.

Velirnirov, A., Plochberger, K., Huspeka, U., and Schott, W. 1992. *Biological Agriculture and Horticulture* 8:325-337.

Scripture Index

Reference	Page	Reference	Page
Genesis 1 (creation)	Preface	Malachi 3:10	6
Genesis 1:4, 10, 12, 18, 21, 25, 31	3	Matthew 4:1-11	Preface, 3
Genesis 1:31	3	Matthew 5:44-45	55
Genesis 1:26	Preface	Matthew 6:1-6	119
Genesis 1:29	130	Matthew 6:3	119
Genesis 1:30	79	Matthew 7:2	8
Genesis 2:7	29	Matthew 7:7-12	7, 119
Genesis 2:8, 15	Preface	Matthew 8:12	Preface
Genesis 2:15	130	Matthew 10:42	7
Genesis 2:16	130	Matthew 20:25-28	9
Genesis 2:17	130	Matthew 22:37-40	Preface
Genesis 2:19-20	3	Matthew 22:39	119
Genesis 3:17-19	130	Matthew 25:31-40	119
Genesis 3:18	3	Mark 4:24	8
Genesis 3:19	130	Mark 12:31	119
Genesis 3:23	130	Luke 6:27-31, 35, 38	8, 119
Genesis 4:2, 12	130	Luke 6:38	126
Exodus 7 to 12 (plagues)	Preface	Luke 10:27	119
Leviticus 19:18	119	John 12:31; 16:11	Preface, 3
Leviticus 25:39-43	121	John 14:30	3
Deuteronomy 15:7-11	121	Acts 20:35	121
Deuteronomy 15:8	7	Romans 8:18-20	3
II Samuel 9:10	130	Romans 13:9	119
I Chronicles 27:26	130	Romans 15:1	120
Nehemiah 10:37	130	II Corinthians 4:4	Preface, 3
Job 1:6-12; 2:1-7	Preface, 3	II Corinthians 9: 6, 8	7
Psalm 17:4	Preface, 3	Galations 5:14	119
Proverbs 1:7	Preface	Galations 6:19-20	6
Proverbs 9:10	Preface	Ephesians 2:2	Preface
Proverbs 19:17	7	Ephesians 4:28	
Proverbs 21:26	7	Ephesians 6:12	3
Proverbs 28:19	130	I Thessalonians 5:14	121
Ecclesiastes 11:1	7	James 2:8	119
Isaiah 14:12	Preface, 3	I Peter 5:8	Preface
Jeremiah 27:11	130	Revelation 21:4	3
Ezekiel 36:9, 24	130		

Index

Endomycorrhizae, 90-91

Energy transfer between species, 119-121

Ericoid, 91

Monotropoid, 91

Orchid, 91

Uptake of many elements, 64, 90-93

Value in phosphorus uptake, 63

Mycorrhizosphere, 90

Nematodes:

In soils, 34-35, 127

Nitrogen release, 78-79

Nutrient release by, 102

Parasitic nature of, 6

Pathogenic, 5, 109-110, 112, 116, 133

Survival after compaction, 107

Nickel, 115

Nimbidin, 110

Nitrifying bacteria, 46

Nitrogen, 9, 10, 20-21, 23, 30, 77-80, 101-102

Accumulation as free amino acids, 116-118

Carbon-Nitrogen ratio, 80-81

Fixation, 76, 93-96, 100

Immobilization, 21, 77-78

In soil air, 42-43

Loss to erosion, 129

Mineralization, 21, 77

Storage in soil peds, 39, 105

Nitrogen Cycle, 77

Nitrogen-fixing organisms, 107

Azospirillum, 10, 95, 100

Azotobacter, 10, 35, 65, 95-96, 100

Bacillus, 96

Beijerinckia, 95

Clostridium, 96

Cyanobacteria, 10, 34-35, 65, 76, 89, 132

Derxia, 95

Enterobacter, 96

Erwinia, 96

Frankia, 95

Herbaspirillum, 95

Klebsiella, 96

Nostoc, 95-96

Rhizobium, 4, 10, 35, 65, 93-95, 100

Nitrogenase, 94

Nutrient interactions, 24

Nucleotides, 114

Organic matter: see Soil organic matter

Organic acids, 45

Oxygen, 9, 10, 18, 22, 25-27, 30, 34, 43-45, 70, 85, 96-100

Root growth, 44, 103-105, 107-108

Oxygen-Ethylene Cycle, 65, 96-100

Anionic nutrient release triggered by, 98

Cationic nutrient release triggered by, 98

Effects of compaction, 107

Ethylene an inhibitor, 97-99

Nitrate an inhibitor, 98

Paramagnetism, 30-31

Parasitism, 3

Character of, 6

Parasitic organisms, 6

Parenchyma, 13-15

Pathogen suppressive soils, 111-112

Pedogenic processes, 55-56

Pesticides, 109, 117, 121, 126

Phenols, 122

Pheromones, 123

Phosphorus, 21, 23, 101-102, 129

Composition in soils, 21

Release in soils, 22, 46, 100

Phospholipids, 67-68

Photosynthesis, 97

Dark phase, 12

CPSIA information can be obtained
at www.ICGtesting.com
Printed in the USA
LVOW04s1342270917
550277LV00008B/92/P